CRISPR PEOPL

T0059856

THE SCIENCE AND ETHICS OF EDITING HUMANS

Henry T. Greely

The MIT Press
Cambridge, Massachusetts
London, England

This book was set in ITC Stone Serif Std and PF DIN by New Best-set Typesetters Ltd. Printed and bound in the United States of America.

Library of Congress Cataloging-in-Publication Data

Names: Greely, Henry T., author.
Title: CRISPR people : the science and ethics of editing humans / Henry T. Greely.
Description: Cambridge, Massachusetts : The MIT Press, [2021] | Includes bibliographical references and index.
Identifiers: LCCN 2020012654 | ISBN 9780262044431 (hardcover)
Subjects: MESH: Gene Editing | Genome, Human
Classification: LCC QH447 | NLM QU 550.5.G47 | DDC 611/ .0181663—dc23
LC record available at https://lccn.loc.gov/2020012654

10 9 8 7 6 5 4 3 2 1

CRISPR PEOPLE

To John and Eleanor, non-CRISPR'd children who taught me about being a parent

Contents

Introduction   ix

Part I: Background   1

1   **Just What Did He Jiankui Do?**   3

2   **Human Germline Genome Editing—What Is It?**   23

3   **CRISPR—What Is It, Why Is It Important, and Who Will Benefit from It?**   33

4   **Ethics Discussions of CRISPR'd Babies before He**   49

5   **The Law of CRISPR'd Babies before He**   75

Part II: The Revelation and Its Aftermath   89

6   **The He Experiment Revealed**   91

7   **The World Reacts—And So Does China**   109

8   **Who Knew What When? Revelations of Pre-Summit Knowledge**   121

Part III: Assessing and Responding to the He Experiment 145

9 Assessing the He Experiment 147

10 Responses 173

Part IV: Human Germline Genome Editing Generally—Now What? 201

11 Is Human Germline Genome Editing Inherently Bad? 203

12 Could Human Germline Genome Editing Sometimes Be Bad? 217

13 Just How Useful Is Human Germline Genome Editing? 225

14 How to Test Human Germline Genome Editing 247

15 The Big Decisions—And How to Make Them 269

Conclusion 293
Acknowledgments 295
Notes 299
Index 371

Introduction

On Sunday night, November 25, 2018, I finished dinner at my home in California and sat down to my computer. As I pulled up my email, a message sent at 7:37 that evening from a friend caught my eye, mainly because of its subject line—"CRISPR babies." The email contained a link to a news story by Antonio Regalado, a reporter with *MIT Technology Review*. The title of the story *more* than caught my eye: "Exclusive: Chinese Scientists Are Creating CRISPR Babies."[1] In it Regalado revealed that a Chinese scientist named He Jiankui[2] was planning to use the DNA editing tool CRISPR (which stands for "clustered regularly interspaced short palindromic repeats") to change the DNA of human embryos that would be transferred into women's uteruses for possible pregnancy and birth—"CRISPR babies."

A quick look at my (exploding) Twitter feed almost immediately led me to an Associated Press (AP) story by Marilynn Marchione. That story was based on at least seven weeks of discussions with He and his colleagues.[3] It confirmed and took further Regalado's story (though without any reference to it), saying nonidentical twin girls had already been born from CRISPR'd human embryos. And just about the same time, He posted a set of videos featuring himself and his colleagues on YouTube, discussing these births.[4]

Dr. He claimed to have overseen the use of CRISPR to modify a gene in the human embryos that is called *CCR5*, a gene known to be important in allowing HIV to infect some human cells. His goal was to make the gene inoperative and thus deprive HIV of that gateway for infection. Two edited embryos, of nonidentical twin sisters, were transferred into their mother's uterus sometime in late March or early April 2018. Sometime in October, somewhere in China, they were born.

Regalado had first posted his article at about 7:15 Eastern Time on Sunday night, but by then it was already after 8:00 on Monday morning in Hong Kong, where the Second International Human Genome Editing Summit ("the Summit") was due to open the following morning. The Summit, which followed a first summit in December 2015, had been organized by the U.S. National Academy of Sciences, the U.S. National Academy of Medicine, the U.K. Royal Society, and the Hong Kong Academy of Sciences. The meeting was a big deal. He Jiankui had fairly recently been added to the list of speakers, though not to talk about CRISPR'd babies. None of the organizers seem to have learned about his actions earlier than the following Thursday; almost all the rest of the world was taken totally by surprise. But his experiment would clearly dominate the Summit and its coverage.

That evening reminded me of another Sunday, more than 20 years before. On Sunday, February 23, 1997, at about 11:00 a.m., my then dean called me at home. I was not accustomed to getting phone calls at home from my dean and started searching my conscience. I did not have long to worry—he said to me, in excited tones, something like "They've cloned a sheep! I thought you should know."

Dolly, the world's first mammal cloned from adult cells, was born in early July 1996, but her birth was kept secret until the researchers, Ian Wilmut and colleagues from the Roslin Institute in Scotland, could publish their scientific paper on her. On the Friday before, *Nature* had sent out its usual press release about its upcoming issue, which included the scientific article on Dolly.[5] Following its usual practice, the story was embargoed until the following Wednesday afternoon, just before the journal's Thursday publication.[6] The distribution of the press release under an embargo is intended to allow journalists time to prepare well-researched stories about an article that will give that article, and the journal, some immediate publicity. Journalists are expected to respect these embargo dates. Instead, the *London Observer*, a British newspaper, ran the story as a front-page exclusive in its Sunday, February 23, edition,[7] and the news flashed immediately around the world.

I don't remember what I said to my dean, but I do remember thinking, "Things are about to get interesting." On Sunday, November 25, 2018, I had the same reaction. Both times, I was right.

This book is my reaction to the He Jiankui "experiment," an experiment that feels like a cross between bad fiction and reckless fiasco, shrouded in a deep fog of secrets. Part I of the book provides some background to the He Jiankui announcement. Its first chapter describes what He actually did, as much as we know, which is still surprisingly little. The second and third chapters explain human germline genome editing and CRISPR. The fourth and fifth chapters describe the ethical discussions about and legal status of human germline genome editing before November 25, 2018.

Part II details the revelation of He Jiankui's experiment in November 2018 in the sixth chapter. The seventh and eighth chapters talk about the fallout from those revelations.

Part III of the book deals with assessments. It weighs the He Jiankui experiment in chapter 9. (Spoiler alert—I conclude it was grossly reckless, irresponsible, immoral, illegal, and probably fattening.) Chapter 10 lays out some immediate responses that the scientific community ("Science") in general should have made, as well as some that China should have done, and describes what has been actually been done.

Part IV, the final part, asks more broadly about human germline genome editing: "Now What?" Chapters 11, 12, and 13 look at the technique broadly, not just in the context of the He experiment. Chapter 11 asks whether the technique is inherently bad (I conclude it is not). Chapter 12 looks at other problems human germline genome editing may cause. And chapter 13 examines the other side of the cost/benefit analysis—would human germline genome editing be very good for anything? Chapter 14 analyzes what kinds of safety testing we would want to apply if we decided seriously to explore going forward with human germline genome editing. And chapter 15 weighs the public policy options of a total ban or regulated use before ending by trying to answer how decisions about this technology should be made. And then, at the end, you get the conclusion.

This book is not entirely second- and thirdhand reporting by me. I was involved importantly in one of the events it discusses and secondarily in some others. I've followed the discussions closely for over five years, and I know personally many of the characters in this play (whether tragedy or farce remains to be seen). All authors will bring their biases and personalities to their work, intentionally or not. In this case, I am trying to give what is, in many places, an openly personal account of the story. You have been warned!

Human germline genome editing invokes many troubling, tricky, and deep general questions. But I am an American lawyer by training. We like looking at cases and working our way from specific examples to broader laws, principles, guidelines, or even rules of thumb. All settings are different, but they provide the opportunity to see consequences and concerns evoked in the real world that might be missed in the solitude of a scholar's study or even in the busy hum of an engaged classroom. The He Jiankui affair puts the many concerns about human germline genome editing—and many other forms of assisted reproduction as well as other human interventions we humans make in ourselves—into a concrete setting. And it tells a fascinating, unnerving, and still unclear story. I hope you enjoy reading about it as much as I enjoyed writing about it.

I Background

1 Just What Did He Jiankui Do?

We don't really know. That's the only honest answer. It is a frustrating answer to give at the beginning of a book about his actions. And it is *not* the answer I expected to be able to give so long after the story first shocked the world. But, as I write this in February 2020, we have far more questions than answers. And the answers we have are from very limited and suspect sources.

We have five kinds of sources. First, and most important, are statements from He Jiankui and his colleagues to the media and in public. Second, there are statements from people outside his immediate research group who were told things by He before, during, and shortly after the news broke. Third, we have some information from talks He gave and comments from people who saw some unpublished manuscripts He had submitted on his preclinical work and his clinical work. Fourth, we have the manuscript of a paper He supposedly submitted for publication.[1] This manuscript, not yet published, was obtained by Antonio Regalado, who wrote about it in early December 2019.[2] Accordinging to Regalado, it was submitted to *Nature* on about November 19, 2018. (I eventually got a copy of it.) And, finally, we have press releases and stories from Xinhua, the Chinese government press agency. One was published on January 21, 2019, and summarized the report of an investigation into the affair by the Guangdong Province. The other reports came out on December 30, 2019, on the day that He and two colleagues were convicted of crimes for their roles in the experiment. Ultimately, the sources for all of those reports are statements or writings by He Jiankui and his colleagues, except, to an unknown extent, the basis for official Chinese statements.

None of these sources should be assumed to be objective, accurate, unbiased, or even honest. The first three sources ultimately go back to He and his colleagues, whose integrity cannot be assumed—particularly in light of evidence of overt dishonesty. The last sources, official Chinese government reports, also must be viewed with suspicion. Any government, whether that of a tightly controlled Communist state or that of a liberal democracy, may well spin facts in ways to make it look better.

It makes me very nervous that, given the time needed to publish books, the earliest you can be reading this book is about a year after I have submitted the manuscript. I will have had some ability to make changes in the following months, but not many opportunities for not many changes. So, we have a mystery of great public interest, with very little information known, and a year's gap between writing and publication. If I weren't afraid that this book may be, at least in part, obsolete before it is published, I would have to be foolish.

And yet—I write in the hope (and confidence?) that this book will not be entirely obsolete. Some facts might (or, given Chinese government control over information, might not) change or be added, but many facts seem unlikely to change. And it is hard for me to imagine new facts that will do much to change the assessments and recommendations in this book. Still, books, like all children (CRISPR'd or not) are hostages to fortune. And, like children, we do our best to help them be good and do well. My deepest anger with He Jiankui is that, in making actual children, he was, in his quest for fame and glory, reckless with their futures. That part of the story is *not* likely to change.

But now to the story itself, at least as we have been told it. In 1984, He Jiankui was born to poor rice-farming parents in Xinhua County,[3] a part of Loudi City,[4] in Hunan province. Xinhua County, a poor region, is near the middle of Hunan province, which itself is a landlocked province in the south-central part of the heavily populated eastern region of China (and the birthplace of Mao Zedong). Xinhua County is about 1,000 miles southwest of Beijing and about 500 miles north and slightly west of Shenzhen, where He ended up working. He's talent must have been recognized early; he was able to attend the best high school in the county: Xinhua No. 1 Middle School 284. Apparently He was obsessed by physics in high school. After being admitted to the University of Science and Technology of China, a highly regarded Chinese university in Hefei, Anhui Province (about 600 miles northeast of his home), he continued to study physics and graduated in physics in 2006.

In 2007, sometime after his college graduation, He received a Chinese national scholarship to study in the United States, moving 7,500 miles this time to start in a Ph.D. program at Rice University in Houston, Texas. According to a laudatory box, entitled "He's on a Hot Streak: Grad Student Jiankui He Scores 3 Major Papers on the Cusp of Earning Doctorate," in a 2010 Rice press release, He applied to three or four graduate programs but was only accepted at Rice.[5] There He received his Ph.D. in biophysics in 2010 in an unusually short time—his dissertation was accepted in November 2010. (Though, in a humanizing note, the Rice press release says that He was president of the Rice

Chinese Students and Scholars Association and in that role organized events for that 400-person community. Also, He apparently loved playing soccer at Rice.)

Professor Michael W. Deem served as He's advisor. Deem is an unusual academic. After earning a bachelor's degree and a Ph.D. in chemical engineering, he did a postdoctoral fellowship in physics. He now holds an endowed chair as the John W. Cox Professor of Biochemical and Genetic Engineering in Rice's bioengineering department but is also a faculty member in Rice's department of physics and astronomy and the founder of Rice's Systems, Synthetic, and Physical Biology program. He is said to be known for his work in "parallel tempering," an improved method for simulating the dynamic properties of physical systems, and in the very different field of improving vaccine development. Deem joined the Rice faculty in 1996, just 11 years before He appeared. Based on his record of honors and positions, Deem appears to be a solid and respected researcher, although perhaps not of the very highest rank (e.g., he is not a member of any of three U.S. National Academies, of Sciences, Engineering, or Medicine).

The Ph.D. thesis by He Jiankui seems to have reflected Deem's diversity of interests. It was entitled "Spontaneous Emergence of Hierarchy in Biological Systems" (and is dedicated to his then-fiancée, Yan Zen).[6] It's a very odd dissertation for someone who would rock the world eight years later by CRISPRing babies. The 213-page dissertation is organized into three main parts: hierarchy in evolving systems (from animal body plans to the world trade network), influenza virus evolution, and bacterial and animal immune systems.[7] The last part includes some discussion of CRISPR not as a tool for humans to use in editing genomes, an idea that was first published by Jennifer Doudna and Emmanuelle Charpentier two years later, but to explain why the front end of the CRISPR construct found in bacteria (as part of their adaptive immune systems) is more diverse than its rear end. Nothing in the entire thesis is at all related to human germline genome editing or to reproduction, human or otherwise.

After finishing his dissertation, He moved to Stanford University, where he spent calendar year 2011 as a postdoc in the laboratory of Professor Stephen Quake. While there, he studied microfluidics, which looks at engineering ways to handle cells and other things in *very* small amounts of liquid. He focused on improving ways to analyze the DNA and RNA of single cells. (I know and am friendly with Quake and probably visited him in his lab during 2011. I have no recollection of, or evidence of, having met He—or any of Quake's students or postdocs then.)

China enticed He to return in 2012, earlier than he had planned, by offering him a spot in the government's prestigious (and financially generous) Thousand Talents program. This program has three parts: one for Chinese professors between 40 and 55 years old, one for foreigners below 55 who are willing to become faculty in China, and one for "young scholars" who must be under 40. The program, the highest academic honor awarded by China's powerful State Council, offers high pay, a prestigious title, special visa privileges, and a one-time bonus of up to 1 million renminbi (RMB; about $140,000).

The subprovincial city of Shenzhen[8] in Guangdong Province on the border of Hong Kong, also contributed to He's return through financial support from its "Peacock Program," an effort started in 2011 to lure talent to Shenzhen by providing subsidies of over $200,000. The sums may not seem large to academics living in Palo Alto, Oxford, or (either) Cambridge, but in China, with its low cost of living (and low cost of hiring staff), they are quite substantial.

In 2012 newly appointed Professor He joined the faculty of the Southern University of Science and Technology, located in Shenzhen about 20 miles from downtown Hong Kong.[9] Back in China, and still only 28 years old, He built a large presence in his university laboratory but perhaps more so in his entrepreneurial activities. Professor He was strongly associated with about eight companies, using, in part, his start-up funds from the Thousand Talents and Peacock programs. His best known firm is Direct Genomics (whose legal name is Shenzhen Bohai Gene Biotechnology Company, Ltd.[10]), which He created in 2012. The company was started to develop inexpensive and accurate single molecule sequencing devices, based on licenses to some microfluidic technologies created by Steve Quake. The firm received more than $4 million in subsidies from Shenzhen and "hundreds of millions of yuan" (said by one source to be about $35 million[11]) from investors. It had a prototype running in 2015[12] and published results demonstrating the effectiveness of its "GenoCare Analyzer" in 2017.[13] In June 2019, after the CRISPR'd babies disclosures, He and Direct Genomics severed all ties.[14]

This is all well and good, you are asking yourself, but what does it have to do with CRISPR'd babies? An excellent question—but aspects of He's background, and of his "nonbackground," are relevant to our story.

At some point, He became intrigued by the possibility of using CRISPR to edit human embryos and, thus, human children. Jennifer Doudna, Emmanuelle Charpentier, and their colleagues published the first article on how humans could use CRISPR as a tool for genome editing at the end of June 2012,[15] and within a year, as more and more scientists published papers on its value, it was clear that this was a major advance. But He, who had written on aspects of CRISPR's use in nature by bacteria at Rice, wasn't doing gene editing. He was doing gene sequencing—reading, not editing.

Then, on April 18, 2015, a group of Chinese scientists at Sun Yat-sen University in Guangzhou published an article in a Chinese journal, announcing that they had successfully edited the ß-globin gene in human embryos.[16] Defects in this gene are responsible for the serious, and not rare, genetic disease beta-thalassemia, but the researchers made clear that their results had been neither efficient nor would have been safe for any embryos used to try to make babies. Few embryos were edited, and many of the edits were in the wrong place. On the other hand, they took no real risks. Not only did they disclaim any intent at transferring the embryos to women's uteruses for possible implantation and pregnancy, they intentionally used human embryos that *could* not lead to a pregnancy. These "tri-pronuclear" embryos had been fertilized by two sperm and, as a result, had three "pronuclei" instead of two, giving them 50 percent more DNA, a surplus that quickly prevents embryonic development.

This work quickly gained worldwide attention, being featured not just in *Science*[17] and *Nature*,[18] but in the *New York Times* (not on the front page, but in a first-section, p. 3 story by Gina Kolata, one of the paper's leading science journalists).[19] Rumors of the work had probably spurred two of the first articles, the previous month, raising concern about such editing (articles that will be discussed in chapter 4.) Did the paper's instant international fame inspire He? It is a question I cannot answer, other than to note that the timing is consistent with this inference.

At least as early as 2016, He began experimenting with CRISPR to edit the embryos of mice, monkeys, and humans. About the same time, he began contacting the laboratories of prominent CRISPR scientists, asking for interviews or lab tours or information. One of his requests, to CRISPR pioneer Jennifer Doudna sometime in late 2016, paid off well.[20] (Doudna says the email came "out of the blue," although there is evidence Doudna and He had met at least in passing before—he attended, but apparently did not present at, an August 2016 conference at Cold Spring Harbor, where he took a selfie of himself with Doudna, an image he posted on his blog.[21])

Doudna and William Hurlbut, a bioethicist who is an adjunct professor at Stanford, had just received a $215,000 grant from the Templeton Foundation to study gene editing. Doudna and Hurlbut had decided to start the grant with a conference at Berkeley in January 2017. In part because they did not have much overseas representation, they invited He to the conference. The results were probably *not* what He had hoped for. As Sharon Begley and Andrew Joseph reported,

> On the workshop's second day, in a session called "Evolution and Human Development," He presented work on using CRISPR to edit mouse, monkey, and human embryos (without pregnancies). His talk did not leave much of an impression, "and I don't think it was received very well," Doudna said. . . .
>
> Worse, another attendee recalled, scientists said He's "science was sloppy and the application unnecessary." One biologist challenged He on technical details of his work, especially how he analyzed the edited genomes for the unintended edits called off-target effects, a critical safety concern. . . .
>
> To CRISPR's leaders, "He wasn't seen as a major player," Doudna said. Having published no papers on CRISPR editing didn't help; neither did presenting research that didn't seem to move beyond what others had reported.[22]

But He got another chance. The prestigious Cold Spring Harbor Laboratory invited him to present that July at its large, three-day Genome Engineering meeting. An early advertisement, though it listed 23 confirmed speakers, did not list He.[23] A later, premeeting website listed He as giving a talk—along with 49 others.[24] That talk, just under 15 minutes long, can be viewed on YouTube.[25] He again presented data on his mouse, monkey, and human work, mainly investigating whether the cells were edited properly (without so-called off-target changes) and whether all the cells were edited and not just some of them. He said that he had injected CRISPR into a human embryo for the first time on November 10, 2016, and had done two or three human embryos most months since then. Begley and Joseph report, "As at Berkeley, the talk left scientists unimpressed. 'It just didn't stand out,' said Doudna, who co-organized the meeting."

The good news for He was that he was meeting all the luminaries of the CRISPR world. The bad news was that they were largely ignoring him. And at some point he decided to try an experiment that, if successful, could not be ignored—to make babies from CRISPR'd human embryos.

In August 2016 He visited Stanford and talked with his former postdoc advisor Steve Quake. Quake says He told him he wanted to be the first to produce genome edited babies, although He gave no indication that he was actively proceeding.[26] Quake says he discouraged him but also pushed him that if he were going to try it, he needed to be sure he had the proper ethics approvals.

But before going forward, He first had to decide what gene to edit. At one point he investigated *PCSK9*, a gene that, in a rare mutation, has been found to confer substantial immunity to coronary artery disease. He had a member of his lab contact Dr. Kiran Musunuru, a researcher at the University of Pennsylvania who had been working for several years[27] on the possibility of using CRISPR in living children and adults to modify the *PCSK9* gene.[28] Musunuru received three emails from FeiFei Cheng, a graduate student in He's laboratory, on November 9 and 15, 2017, and again in January 2018. Musunuru provided little advice and no encouragement to the student. The emails were seeking advice about trying to demonstrate whether using CRISPR to knock out the gene in humans was a reasonable and feasible approach and, if so, how to do it. They did not mention any plan to make human babies.

As with so many other questions in this saga, we do not know whether He used CRISPR to modify the *PCSK9* gene in any of the human embryos he was preparing for possible transfer, implantation, and birth. One article reports that He explored *PCSK9* first but then switched to *CCR5*:

> Dr. He was investigating editing a gene that can offer protection from familial hypercholesterolemia, a rare cholesterol-related disease that can cause broken bones in children. He changed his mind after visiting a village where he saw HIV-positive families facing discrimination, people close to him say. Children born to infected individuals weren't able to attend regular schools. He saw a gene-editing trial as a way to use science against that injustice.[29]

The gene, *CCR5*, provides the template for the body to create the CCR5 protein. This protein is found on the outer surfaces of the membranes of several types of human cells, including, notably, T cells, a crucial part of the immune system and the part that is most ravaged by HIV infection. Many forms of the HIV virus can only invade T cells if the T cells carry on their surface both a CCR5 protein and another protein called CD4. T cells without a CCR5 molecule should be more resistant to HIV infection. And, in fact, there is some human evidence for this.

Some humans have natural mutations in one or both copies (one copy from their mother, one from their father) of their *CCR5* genes that prevent the protein from being made properly. About 10 percent of Northern Europeans have one nonfunctional copy of the gene; about 1 percent have two nonfunctional copies and make *no* CCR5. These nonfunctional copies almost always have the same mutation—they are missing the same 32 base pairs ("letters" of DNA) from an important part of the gene. DNA "spells" out the instructions for making protein in words of exactly three letters. If you add or subtract a number of consecutive base pairs that is not evenly divisible by three, such as 32, you not only miss out on some of the words (10⅔ in this case), but you change all subsequent words. If a stretch had said ACG TAG GAA TTA and you delete the first AC, it now reads GTA GGA ATT A. Different letters, different words, a different protein, and often a truncated one, as one of the "new" words may well say "stop." Three different three-base pair combinations ("words") literally tell the cell to stop making protein.

Humans with two copies of this 32-base-pair deletion in *CCR5* are known to be less likely to become HIV infected. The possibility of providing this resistance in babies through genome editing had been raised before. *CCR5* is one of 10 genes on a list that Harvard scientist George Church was showing as early as 2015 of genes with rare variants that strongly protected against various diseases.[30] At some point no later than March 2017, He decided to try to knock out that gene in embryos in order to provide the resulting children with protection from HIV infection. (His mouse, monkey, and human work in editing embryos had begun in 2016; his submitted manuscript says he made *CCR5* edits in mice, monkeys, nonviable human embryos, and human embryonic stem cells, but I cannot determine when, other than "before initiating the clinical trial.")

At least some people who heard him talk about it think He's interest in *CCR5* and HIV/AIDS was sincere. As He has said, HIV/AIDS is a real problem in China. The prevalence of HIV infection is much lower than in most of the West, but it is not trivial. People who are HIV positive are greatly stigmatized. Hunan, He's home province, is one of the more heavily affected areas of China.

By the time his lab was contacting Musunuru, He had already decided to use *CCR5*. He Jiankui needed to find HIV-positive families because on March 7, 2017, only five weeks after the workshop at Berkeley gathering, "He submitted a medical ethics approval application to the Shenzhen HarMoniCare Women and Children's Hospital that outlined the planned *CCR5* edit of human embryos."[31] According to the manuscript of the scientific publication he supposedly submitted for publication on this work, he received permission in that month from a hospital ethics board to commence his experiment.[32] (Steve Quake says he first learned of the ethics committee approval in an email from He in June 2017.)

By June 10, 2017, He was actively recruiting prospective parents for his trial. An excellent article in *Science* starts,

> On 10 June 2017, a sunny and hot Saturday in Shenzhen, China, two couples came to the Southern University of Science and Technology (SUSTech) to discuss whether they would participate in a medical experiment that no researcher had ever dared to conduct.[33]

Sometime before then—we don't know when—He had contacted an HIV/AIDS support group called Baihualin China League and sought their support to find couples who would be willing to volunteer to try to have a "genetically" HIV-resistant baby. The group's founder reportedly introduced about 50 families to He's team (something he now regrets).[34] He's goal was to find a particular kind of HIV family, one where the father was infected and the mother was not.

By September, eight married couples had agreed to participate; one pair subsequently withdrew. According to the consent forms used and other sources, the families also received free fertility treatments, medical care for the pregnancy, and a stipend. The total value of the benefits for those whose babies were born has been stated, based on the Chinese consent form, as the equivalent of about $40,000.[35]

Eventually, five women had a total of 13 embryos transferred for implantation.[36] The first attempt was made no later than January 2018, but it failed. Sometime in late March or very early April 2018, He succeeded. Two female embryos from one couple were transferred into the wife's uterus. According to He's manuscript, they harvested 12 eggs from the babies' mother and injected them after fertilization with guide RNA and Cas9. The fertilized eggs yielded four blastocysts (five- to six-day embryos); two had been edited, two had not been. The two edited blastocysts were transferred into the mother's uterus for possible implantation. (Many months later, sometime in the early fall, He succeeded in creating another pregnancy from a genome edited embryo.)

Ten days after the transfer, He sent Quake an email, with the subject line "Success!," saying that "the embryo with CCR5 gene edited was transplanted to the women [sic] 10 days ago, and today the pregnancy is confirmed!"[37] (As detailed in chapter 7, he later told a few other people of this pregnancy.)

Then, sometime in mid-October (we still don't know the date), He emailed Quake that the babies had been born. According to the *Wall Street Journal*, "One October evening, the twins' expectant father called a member of Dr. He's lab to say his wife was going into labor. Dr. He raced to Shenzhen airport, postdoctoral students in tow, and flew north."[38] (Interestingly, the article He submitted to a major scientific journal says, twice, that the babies were born in November, which is inconsistent with the other reports. Normally, I might trust the paper's author, but not in this case.)

The woman gave birth that night by emergency cesarean section to nonidentical twin girls. In his public statements, He referred to them using the pseudonyms Nana and Lulu. The site of the birth has not been disclosed, although the fact that Dr. He, located on the south coast of China, flew north for the birth is at a least some slight clue. The emergency delivery was probably because Nana and Lulu were born too soon—with the pregnancy achieved in late March, it would have gone only about 30 or 31 weeks and the delivery would have been quite premature (which is not unusual for twins). Whether this was caused by premature labor, fetal distress, or maternal distress is not known. According to He's late November statements, the girls were healthy by late November when He spoke about them onstage, although in late October or early November, Ryan Ferrell, He's public relations expert, asked Quake's help in convincing He to delay the announcement in part because "the twins are still in the hospital, so no positive imagery." The He manuscript does not mention any prematurity but says,

> The women [sic] delivered normal, healthy twin girls, named Lulu and Nana, in November 2018 in China. The Apgar scores which measures [sic] skin color, pulse rate, reflexes, irritability grimace, activity, and respiratory effort were 8 points for Lulu and 9 points for Nana, out of a maximum of 10 points.[39]

Onstage at the Second International Summit in Hong Kong, He responded to a pressed question about other pregnancies by saying a second woman had also become pregnant but that that pregnancy was at a very early stage. We had been told by the Chinese news agency that the second pregnancy was still ongoing in late January 2018, but until December 30, 2019, we did not know the result. According to the reports from He's trial, that baby was also born—although we do not know any further details, including notably when the baby was born, what gene was edited, how successful the editing was, whether the baby is healthy, or really, anything other than his or her existence. From the calendar, though, we know that even a pregnancy that was in its very early stages in late November 2018 could not have lasted beyond perhaps early August 2019.

What else do we know about the experiment? We know something about the genomic results of the experiment from He's presentation at the Hong Kong Summit, a newspaper story he cooperated with that was being prepared for at least several weeks before the Summit, five YouTube videos he released just before the Summit, and from the manuscript discussing the experiment that He allegedly submitted to a medical scientific journal, entitled "Birth of Twins After Genome Editing for HIV Resistance." (He has not published any peer reviewed publications at this point on his human experiment or on his nonhuman CRISPR editing work.)

Based on He's slides from his Summit presentation, as carefully pored over by researchers through still shots of the presentation's video recording, neither twin's DNA was successfully edited to carry the sought-after 32-base-pair deletion.[40] This again is inconsistent with the abstract of the manuscript submitted by He, which says "We used CRISPR-Cas9 to reproduce a prevalent genetic variant of the *CCR5* gene in fertilized oocytes during *in vitro* fertilization procedure." The text of the article, however, does make clear that neither of the twins had the 32-base-pair deletion that is the "prevalent genetic variant" of *CCR5*.

One twin has two *CCR5* genes with "frameshift" changes (deletions of a number of base pairs not evenly divisible by three) that should make them nonfunctional, although neither copy is edited in a way ever seen before in humans. The other twin, however, is more complicated. The copy of *CCR5* on one of its chromosomes is normal. The other has a deletion, but it is a 15 base pair deletion. That 15 is important because it is evenly divisible by three. Depending on where the deletion starts (at the end of a "codon," a three-letter "DNA word," or in the middle of one), it may or may not have caused a frameshift. If the deletion starts at the end of a codon, it just deleted five amino acids from the eventual CCR5 protein, which may or may not make that protein inactive. If the deletion starts inside a codon, it would be a frameshift. But, whatever the state of the edited copy of the gene, one copy is not edited. This means that all of her T cells should have a normal copy of the CCR5 protein, though perhaps in smaller quantity than in people with two functional copies. Her T cells may have some increased resistance to HIV infection, but it should not be strong.

And that is nearly all we know. But we do have two other sources of information about the experiment.

On January 21, 2019, the Xinhua News Service, the press agency of the Chinese government, published a 316-word article in English that reported on the results of an investigation of He's experiment by authorities in Guangdong Province, where the work had taken place.[41] It is worth copying the entire Xinhua press release:

GUANGZHOU, Jan. 21 (Xinhua)—A preliminary investigation into the claimed "genetically edited babies" shows that Chinese researcher He Jiankui had defied government bans and conducted the research in the pursuit of personal fame and gain.

The investigation team of Guangdong Province announced on Monday that He had intentionally dodged supervision, raised funds and organized researchers on his own to carry out the human embryo gene-editing intended for reproduction, which is explicitly banned by relevant regulations.

He Jiankui, associate professor with Shenzhen-based Southern University of Science and Technology, claimed in November 2018 that the world's first genetically edited babies were born, and their DNA was altered to prevent them from contracting HIV.

According to the investigation, in June 2016, He started the project and organized a team that included some overseas members. He conducted the gene-editing activities using technologies without safety and effectiveness guarantee.

With a fake ethical review certificate, He recruited eight volunteer couples (the males tested positive for the HIV antibody, females tested negative for the HIV antibody) and carried out experiments from March 2017 to November 2018.

As HIV carriers are not allowed to have assisted reproduction, He asked others to replace the volunteers to take blood tests and asked researchers to edit genes on human embryos and implant them into the females' body.

Two volunteers were pregnant. One gave birth to twin girls Lulu and Nana. The other is still pregnant. One couple quit the experiment halfway through, and the other five couples did not conceive.

The activities seriously violated ethical principles and scientific integrity and breached relevant regulations of China, according to the investigation.

Officials in charge of the investigation said, He, as well as other relevant personnel and organizations, will receive punishment according to laws and regulations. Those who are suspected of committing crimes will be transferred to the public security department.

The babies and the pregnant volunteer will receive medical observation and follow-up visits.

And then, over 11 months later, the Chinese news services released several articles about the trial, conviction, and sentencing of He Jiankui and two of his associates. Xinhua released at least two stories plus a "perspective" on the trial, all dated the same day, December 30, as the trial. One of the news releases, worth quoting almost in full, states,

In accordance with a ruling handed down by Nanshan District People's Court of Shenzhen City, He was sentenced to three years in prison and fined 3 million yuan (about 430,000 U.S. dollars) for illegally carrying out human embryo gene-editing intended for reproduction, in which three genetically edited babies were born.

He used to be an associate professor with the Southern University of Science and Technology. Zhang Renli and Qin Jinzhou from two medical institutes in Guangdong Province received jail terms of two years and 18 months with a two-year reprieve, respectively, as well as fines, said the Nanshan District People's Court in a verdict.

Public prosecutors said that the three, who were not qualified to work as medical doctors, had knowingly violated the country's regulations and ethical principles to conduct gene editing in assisted reproductive medicine.

The prosecutors showed substantial evidence to prove He's team fabricated an ethical review certificate and recruited eight volunteer couples (with men who tested positive for HIV) intending to produce HIV-immune babies. They implanted genetically-engineered embryos into the women's body and impregnated two of them, who gave birth to three babies.

The three, whose acts were "in the pursuit of personal fame and gain" and have seriously "disrupted medical order," should be punished, the court declared.

The three pleaded guilty during the trial.[42]

These two Chinese sources, from January and late December 2019, confirm that babies exist (two in January, three by

December) and (apparently) that their DNA was edited. Given the Chinese government's impressive ability to control the flow of information (and of people), we may never know more about what actually happened. But this book is not finished. We do know enough about the background, the way the information came out, and the experiment's implications that the book has more stories to tell and much more to say.

2 Human Germline Genome Editing—What Is It?

Human germline genome editing is what He Jiankui did—and the subject of this book.

Bioscience is not good at English. Sure, "English" is now the international language of science; any scientist, whatever his or her home country, who does not publish in English runs a very high risk of irrelevance. The same is true of any scientist who cannot give a presentation in a mix of PowerPoint and English. But the "English" of bioscience is too often a nasty mix of crypto-Orwellian acronyms ("COAs"[1]), Latinate neologisms, and just plain weirdness (like the "sonic hedgehog" protein). Plain English is rare.

The method He used in his experiment, the demonstration of which was the point of the experiment, is usually expressed in "plain English." The problem is both that several different plain English phrases are used for it *and* that, when used precisely, the words of those phrases turn out not to be so plain. In this book I will use the term "human germline genome editing" (or sometimes just "germline genome editing"). I want to start this chapter by explaining exactly what that phrase means or, at least, what *I* mean by it. I will then turn to why the method it describes is so very controversial.

"Human Germline Genome Editing"

Let's start with "human." This clearly denotes something that is not a mouse, monkey, worm, redwood tree, or, for that matter, a rock. But here it has a more limited meaning. "Human" here could just mean that the DNA being modified is (ultimately) from a human being, but I am using it to mean that human DNA is being modified in an effort to create a "human person," or, more explicitly, a baby. Although much important research is done editing isolated and purified human DNA or DNA in human cells, this book is concerned with whole people, not their bits and pieces.

Some people argue that human embryos are whole people. Without debating that here, I will just declare that this book is not about editing DNA in human embryos that are intended only for research and will never be in a woman's uterus—and hence will never have a chance to become a baby. Different people assign different moral values to human embryos that are *ex vivo* (outside the living, also called *in vitro*, for in glass, even though they are actually in plastic); what is important in this book is that He's experiment was on embryos that he intended to turn into babies.

Then what is a "germline"? As far as I know, all multicellular organisms have specialized cells, cells that perform particular functions that are different from those performed by some of the organism's other cells—our skin cells do different things than do our heart cells. Starting with August Weismann in 1892,[2] biologists have classified the cells of most multicellular organisms as one of two kinds: somatic cells or germ cells. Somatic cells make up the body (the meaning of the Greek word "soma"), and they live and die with and as part of the body. But almost all multicellular organisms also have other cells, called the germ cells or the germline, that can, in effect, survive the organism's death by creating the next generation. ("Germ," unlike our usual association of it with invisible disease-causing microbes, comes from a Latin word for bud, shoot, or sprig, which, in medieval France, took on the meaning "seed.") In humans, the germ cells give rise to eggs and sperm.

It turns out, though, that reproduction in the biosphere is often quite different from what we expect, based on our experience with ourselves, other mammals, and other vertebrates (mammals, birds, reptiles, amphibians, and fish). The distinction between somatic and germline is meaningless in single-celled organisms, which make up the vast majority of the Earth's species and individual living organisms. These cells, which reproduce by dividing in two, are either never germ cells or always germ cells. A truly single-celled organism cannot have any specialized cells, although a few come somewhat close. Slime molds, for example, are single-celled organisms that can come together and cooperate in a multicellular mass to form a "fruiting body" for purposes of reproduction—but that is mainly useful as one more example that all rules in biology have exceptions.

And even some multicellular organisms can "reproduce" without using specialized cells. Sponges, for example, sometimes reproduce through "fragmentation"—some cells just fall away from the sponge and can produce new (though genetically identical) sponges. It is said that some fishery workers, when they capture starfish that are eating their clams or oysters, have cut them in half and thrown them back into the water. This will usually just lead to two starfish instead of one as each half will regrow its missing side. Similarly, some animals, notably some sea anemones, will just split themselves in half, reproducing by fission.

Some other organisms can reproduce through, in effect, budding. A new animal grows from the edge of the old. Sometimes, as with hydra, it eventually separates; sometimes, as in some corals, they stay attached. Many kinds of plants reproduce through something akin to budding. New stems or trunks grow up from roots or runners underground to form new copies of the plant; they may stay connected or may eventually end up separating. Arguably the world's largest living organism is a stand of aspen trees, named "Pando," that includes 47,000 trunks spread over 106 acres near Richland, Utah.[3]

I put "reproduce" in quotation marks above because the new copies produced in these ways are, except for possible mutations during their development, genetically identical to their progenitors. In some cases they remain connected to their progenitors. Is this reproduction or something else? Whatever it is, it does not involve germ cells or a germline.

At the same time, other organisms reproduce in ways that are strange to us but still involve a germline. Some species, such as rotifers, use forms of cloning that start with germ cells. In those cases eggs develop as identical copies of their "mothers." Those eggs are part of their germline. (Some rotifer species, interestingly, can switch back and forth between this kind of cloning and sexual reproduction.) Clonal rotifer reproduction is a form of parthenogenesis (virgin creation) because no males and no sperm are involved. Other forms of parthenogenesis exist wherein only the female cells are used, but these produce only partial clones—the egg cells go through the process of meiosis, wherein they lose one of each of their pairs of chromosomes and then two of them recombine. Some species of sharks, frogs, salamanders, snakes, and lizards regularly reproduce by parthenogenesis. This isn't your parents' sexual reproduction, but it too relies on a germline.

Change the *CCR5* genes in a T cell—a somatic cell that is a white blood cell and part of a human's immune system—and that change will live and die with that T cell and its progeny. It does not move into other cells in the human, and it dies when that T cell and its descendants die, no later than the death of the person who carries them. Change the *CCR5* genes in fertilized eggs and, if those fertilized eggs become babies, you have changed the *CCR5* genes in every cell of those people, from brain cells to heart cells to liver cells to . . . their eggs and sperm. This means that the *CCR5* edit could survive those people and might be passed on to their children, grandchildren, great-grandchildren, and so on until the end of their line of descendants. *That* is germline editing.

Note that this may not affect all those endlessly forward-stretching future generations. Perhaps the edited person will have no children, or have children with no children, or grandchildren with no children. Or, even if the line continues, the people in it may not have the edited copy of the gene. Even if both copies of the initial person's *CCR5* were successfully edited, the children will get only one *CCR5* copy from that parent; the other one will come from their other parent, who presumably has two "normal" copies of the gene. The children would then have one edited copy and one unedited copy; half the time their children will inherit the edited version, but half the time they will inherit their other parent's unedited version. Unless it confers a strong benefit on those who carry it, the edited version of the gene might well disappear, either quickly or by slowly trickling out. And, of course, the edited version itself could disappear because of subsequent natural mutations or future "reediting."

Still, this possibility (not certainty) that the descendants' genes will be altered is what most upsets many people when it comes to editing the genes of early-stage embryos. Potentially, at least, one could change genes of the human species forever. That's unlikely but not impossible—and, to many people, very disconcerting.

People have much less trouble with "somatic cell genome editing." This edits a person's genes during that person's life but only in particular tissues or organs, usually to treat or prevent disease. As long as the edits do not affect the person's eggs or sperm, the germline is unaffected, as is the DNA of future generations. This concept, called "gene therapy,"[4] is genome (or gene) editing but of the somatic cells rather than the germ cells: human somatic genome editing, rather than human germline genome editing. A few versions of it have already been approved and are in clinical use for particular diseases.[5] It is likely to be the most important use of human genome editing, but, because it lives and dies with the patient, it is much less controversial—and so is not this book's concern.

But between adults—or even fetuses—having DNA in their livers or hearts edited, on the one hand, and a single-celled fertilized egg (called a "zygote"), on the other, sits the embryo. After the fertilized egg splits, an embryo is made up of more than one cell. What happens if the edits affect some cells but not others?[6] The result is called a "mosaic," like the walls or floors made up of different colored small bits of glass or tile. Some cells in that embryo might have two copies of an edited *CCR5* gene, and some will have no edited copies of the gene. (And some, if the editing was incomplete, may have only one edited copy.)

For some time after fertilization, the embryonic cells that may become a person's body can become any type of body cell, but that changes quickly as the cells begin to specialize or to form precursor cells that will form heart cells, brain cells, bone cells, and so forth. As early as the 12th to 14th day after fertilization, some of an embryo's cells start to specialize to form a line of cells that will eventually produce, among other things, the germline and, ultimately, eggs and sperm. One might, in theory, intentionally do genome editing at, say, day 20, in such a way that only the cells that cannot become germ cells are edited—or so that only the cells that can become germ cells are edited. Or, if the earlier gene editing produced a mosaic embryo, the same result could come about by accident. All (or some) of the cells that eventually became the germ cells might have, or not have, the edited *CCR5* gene. So this embryo editing might, or might not, be germline editing.

Ultimately, this is why I write of *germline* editing rather than embryo editing. If you edit a zygote or edit all the cells of an early embryo, you are necessarily editing both somatic cells and germline cells (by editing cells that will eventually give rise to both of those categories). That is why zygote editing is necessarily, and early-stage embryo editing is usually, germline editing. In a later-stage embryo, though, one might edit only the cells that have already been differentiated along a path that means they could not become germline cells. That would be embryo editing, but not germline editing.

On the other hand, you don't need to start at the embryo to edit human germline genomes. One might take egg and sperm samples from people and edit their *CCR5* genes while the cells sit in a petri dish in the laboratory (in vitro, even though it is plastic, not glass). One could then use these edited gametes to create an embryo, rather than edit the embryo itself.[7]

One could also edit the germline by putting the editing mechanism into living babies, adolescents, or adults (in vivo). If you edit the genome of the eggs of 20-year-old women or the sperm-forming cells of 20-year-old men, you edit their germline.[8] (At least one group has already suggested in vivo germline genome editing by doing "gene therapy" on sperm through an injection into a man's testicles.[9] I'm not sure how many volunteers that would get.) This might be done on purpose, by injecting the editing agent into the ovaries or testes. But it might also happen inadvertently, by providing gene therapy to a person with the goal of affecting a particular organ but unintentionally, and possibly unknowingly, also editing the person's eggs and sperm.

Logically, these options are also human germline genome editing, whether intentional or accidental. But they are not what He Jiankui did or what people are currently worried about others repeating. It is worth remembering that it can happen, but in this book, when I talk of "human germline genome editing," I do not include intentional or accidental editing of eggs and sperm.

I use the term "genome" editing and not "gene" editing or "genetic" editing because such modifications may change more than one gene or change DNA that is not in what is usually considered a "gene." The "genome" is all of the DNA sequence in a person—the 6.4 billion base pairs (the adenines, cytosines, guanines, and thymines—As, Cs, Gs, and Ts) in the 46 chromosomes found in the cell's nucleus as well as the 16,000 or so base pairs found in the DNA of the cells' mitochondria. To be sure, the particular example of He's experiment—editing *CCR5*—targeted one specific gene and could be considered "gene editing" or "genetic editing." Many potential uses for germline editing—especially for enhancement, the most controversial use—will no doubt require changes to many pieces of DNA, some in genes and some not.

It is tempting just to call it "DNA editing"—after all, editing DNA is fundamentally what it does. I prefer "genome editing" because it reinforces the idea that this is not just editing a piece of DNA in a laboratory but editing the whole information-storing assembly that a person's DNA makes up.

And, finally, I use "editing" instead of "modifying" or "changing" or "mutating" (just a fancy Latinate word for changing) for two reasons. First, editing is a technique used to make a specific intentional change: for example, to change an A to a G, to delete a particular stretch of 32 bases, or to add an important 12-base sequence. Each of the other words accurately describes what is done, but they do not carry the same sense of intentionality. Note that this is intention, not necessarily result. He Jiankui intended to edit *CCR5* genes by deleting a specific stretch of 32 base pairs. That he failed does not mean he did not do human germline genome editing, just that he did it without complete success.

Second, although this book's title includes "CRISPR," human germline genome editing need not be "CRISPRing." Although He Jiankui says he used CRISPR, not all gene edits today must use CRISPR. Other, older methods, though more expensive, time-consuming, and difficult, can also be used to edit DNA. Transcription activator-like effector nucleases (TALENs) and zinc finger nucleases (ZnFs) are the two most significant. (Oddly enough, the first one is mainly known by its acronym while the second is often referred to by its full name, perhaps because the first name is hard to remember and the second is easy.) For the near future, CRISPR seems likely to be used much more often, but, also important, refinements upon CRISPR have already begun to appear. Replacements for it are likely, with different sets of pluses and minuses. When referring to CRISPR as a specific technique, I may use "germline CRISPR" or even "CRISPR'd babies," but the general topic is "genome editing," not "CRISPR."

That's what I, and this book, mean by "human germline genome editing." Lots of variations on "human germline genome editing" appear in scientific publications and the popular press. Sometimes you will read about "human genome editing" rather than "germline genome editing." That is a fair use if the user intends to include somatic cell editing *and* germline editing, as in the report by the U.S. National Academies of Sciences and of Medicine, and in the Hong Kong Summit that led to the revelation of He's experiment. You may read about human germline "gene editing," human germline "genome editing," or human germline "DNA editing"—and often this won't include the important qualifiers "human" and "germline." This is usually just a shortcut that saves writers, especially headline writers, a few characters. And, finally, you may read of human germline genome (or gene or DNA) "CRISPRing" because that is the best technology at our disposal today. But not, perhaps, tomorrow. Usually, however, whatever words they use, articles about the He Jiankui experiment and its implications are talking about what I'll call "human germline genome editing": making intentional changes to the genomes of embryos that, it is hoped, will become people, people whose eggs or sperm will carry those changes.

3 CRISPR—What Is It, Why Is It Important, and Who Will Benefit from It?

It is a tool, and one that leapt far beyond the existing tools—perhaps not as far as a chain saw leapt from a stone ax, but close. And we should all benefit from it, though which scientists, universities, and companies will particularly benefit remains to be seen.

As noted earlier, CRISPR is also an acronym, standing for "clustered regularly interspaced short palindromic repeats."[1] This is a case where my distaste for COAs gives way. Not only is CRISPR a much shorter and easier to remember name for this tool but it makes a catchy word, one that can make a good sounding noun, but also a verb, and one that combines nicely with "-ed" and "-ing." I love the name CRISPR, although, somewhat to my surprise, someone else has published an article on how and why he hates it (mainly because there is nothing crispy about it).[2]

And in another, earliest sense, CRISPR refers to molecular constructs that bacteria have used to defend themselves from viral invaders. In this sense, CRISPR goes back billions of years—perhaps three billion or more. The first time humans noticed them appears to have been in 1987, when a Japanese group published the sequence of one gene, along with some surrounding DNA sequences, of the ubiquitous gut (and laboratory) bacterium

E. coli. They found something never seen before—DNA that had a set of identical 29 base pair sequences, separated from each other by four sets of 32 base pairs, which they called spacers, each with a different sequence. They had no idea what this DNA did and wrote "the biological significance of these sequences is not known."[3]

In the early 1990s other scientists, most notably Spanish researcher Francisco Mojica at the University of Alicante, saw odd debris inside bacterial cells. In trying to understand them, he ran into this same kind of spacer arrangement. Mojica pushed to understand this phenomenon, publishing several articles on its mechanism as he understood more about it.

Naming of CRISPR

Others also were investigating the same repeating bits of DNA inside bacterial cells, but the phenomenon had no accepted name. In 2000, Mojica published an article on its biological significance, calling it "SRSR" for short regularly spaced repeats.[4] (Mojica has pointed out that this acronym nicely mirrored the way the clusters were organized: spacer/repeat/spacer/repeat.[5]) In 2002, Ruud Jansen in the Netherlands published an article on the same phenomenon, which he called "spacers interspersed direct repeats," or SPIDR.[6]

Mojica says,

> The potential naming conflict was solved after mutual agreement of the two research groups to use CRISPR (pronounced krisper) after "Clustered Regularly Interspaced Short Palindromic Repeats." Jansen immediately accepted the new definition and acronym rather than the other, less descriptive or not so distinctive alternatives that were proposed.[7]

The term seems to have been used first in print by Jansen, who wrote later in 2002:

> To acknowledge the joining of this class of repeats as one family and to avoid confusing nomenclature, Mojica *et al.* and our research group have agreed to use in this report and future publication the acronym CRISPR, which reflects the characteristic features of this family of clustered regularly interspaced short palindromic repeats.[8]

(The article's acknowledgments section states, "We enjoyed pleasant discussions with Francisco Mojica of the University of Alicante, Spain, about the renaming of CRISPRs.")

Nailing down from publications the origin of the term CRISPR was satisfying, but it did not illuminate a question I really wanted answered: Were Mojica or Jansen influenced by the good sound of "CRISPR," at least in English? So I emailed them. To my pleasant surprise both answered me. Mojica responded very quickly, in a very friendly and helpful email.[9] In his memory, from 18 years ago, Jansen had proposed that, to avoid the confusion of two different terms, the two of them should rename these families of repeats to something they (and, he hoped, others) would use consistently. Jansen asked Mojica for suggestions.

Mojica had two goals for the name: he wanted the name to reflect the importantly distinctive feature of this family of repeats, and he wanted it to be easy to pronounce, especially in English. He also realized that they did not know enough about the function of CRISPR to give it a name that implied any function. So he proposed "regularly spaced short repeats" (RISR), his own favorite, and CRISPR. Jansen also said he preferred CRISPR and noted that "Also not unimportant is the fact that in Med-Line CRISPR is a unique entry, which is not true of some of the other shorter acronyms."[10]

As to the variations on CRISPR, Jansen told me, "At the time of the publication I could not imagine that there would ever be the verbs crispr, crisprs, crispring and crispred, but it sounds very well."[11] Mojica said he had not thought about how well CRISPR could be used as a verb; it is 'just a lucky coincidence.'" Mojica added

> as it is also the fact that one of the meanings of the word Crisper, with identical pronunciation, is "a compartment *for storing* fruit and vegetables" and CRISPR is a segment of DNA *for storing* chunks of invading genetic elements: both are storing devices. [emphasis in original][12]

Jansen also told me that the sound of CRISPR resembles the Dutch verb "knisperen." This means to "crisp" as, Jansen said, "the freshly fallen autumn leaves or the crust of a freshly baked croissant. With these connotations, you can understand that CRISPR was the acronym of choice."[13] Prompted by this email, Mojica pointed out that

> In Spanish, the closest word to CRISPR is the verb crispar (to annoy in English), pronounced *krees pahr*. In consequence, many Spanish people pronounce CRISPR that way, not at all appropriate in the context of the CRISPR field. Even worse, much more people say "krispies," which really annoys me. Yet, I believe we made a great choice.[14]

Jansen summed up his experience by telling me,

> At the time of publication I changed jobs and quit the work on the CRISPRS, because we could not get any funding for our following CRISPR project. I considered it very likely that the article would disappear in the pile of forgotten publications, although I hoped somebody would be able to solve the mystery of these abundantly present prokaryotic genetic units. The rest is history and I am happy to be part in the planting of this very vital seedling.[15]

As someone who has had some say in naming two children, the discussions between Mojica and Jansen, though very different in some details, have a very familiar feeling. And it is, indeed, quite a child they named.

Around 2005, Mojica and others realized that CRISPR was actually a form of an immune system for bacteria and archaea (another, nonbacterial, type of microbe[16]), and not just any immune system but one similar, in a key way, to the major human immune system, the adaptive immune system.[17] This was an impressive discovery; no specifically bacterial (or archaeal) immune systems had been identified before. Humans and other animals use many different ways to defend themselves against microbial invaders, but one category of defense mechanism recognizes specific invaders that it has seen before and attacks only them. This system is the basis for vaccination, as well as for your immunity to some infectious diseases, such as measles, if you have had them before. CRISPR does the same kind of thing with viruses, which don't just plague us during cold and flu season, but also attack bacteria and archaea. (Viruses that attack these microbes are generally called "bacteriophages" or just "phages," from the Greek word "to eat.")

Microbes with a CRISPR system, if they successfully fight off a virus's attack, will cut up the virus's DNA (and also, it seems, the RNA in RNA-based viruses, but we'll ignore these) and put bits of the virus's sequence into their own DNA. It is these pieces of viral DNA incorporated into the microbe's own DNA that are the "clustered regularly interspaced short palindromic repeats"—CRISPR. No need to worry about any of the specific words in that name; it's sufficient to know that the DNA in a CRISPR region will be used as a template for RNA (which is most of what DNA does).

Basically, the CRISPR section of the microbe's genome creates a homing mechanism. It serves as the template for the microbe to make an RNA molecule based on the CRISPR DNA template. The template is based on one of crucial aspects of the secret of life—how DNA, RNA, and proteins are related. The DNA bases, A, C, G, and T (adenine, cytosine, guanine, and thymine), bind to each other only in very specific ways. A binds with T, and T with A; C binds with G, and G with C.

With CRISPR, a bacterium will make an RNA molecule from the CRISPR template. The RNA molecule will have As where the DNA has Ts and Gs where it has Cs. In a complication unimportant for our purposes, RNA uses a U (for uracil) instead of a T. So, if a CRISPR section of the microbe's DNA reads ATTTGGCAC (i.e., a sequence that it found in a past viral invader) the microbe will make an RNA that reads UAAACCGUG. This is called the "guide RNA." The guide RNA (made from the template in the bacterium's DNA) will diffuse through the bacterium's cell.

Drifting along with, and attached to, the guide RNA is a microbial enzyme, a kind of protein. If, floating in the bacterium, there is viral DNA containing the complement of the guide RNA's sequence, in this case ATTTGGCAC, the guide RNA will stick to it like glue, its bases holding on to their complements in the viral DNA. (You have, I hope, noticed that this is the same sequence we started out with, the old virus's DNA sequence found in the CRISPR region.) The attached protein, like a molecular scissors, cuts this viral DNA into pieces at the location the guide RNA has found. Unlike RNA, DNA comes in two connected strands of sequence, curling around each other in the famous "double helix." (Think of a twisted ladder.) The associated protein cuts the viral DNA across both strands, "killing" the virus, making it unable to reproduce, and thus protecting the microbe.[18] The first of these "scissors" proteins was given the poetic name "CAS," for "CRISPR-associated" protein. It turned out to be one of a large family of CAS proteins and so had its name changed to Cas1. The great science writer Carl Zimmer summarized the whole process better than I can:

> As the CRISPR region fills with virus DNA, it becomes a molecular most-wanted gallery, representing the enemies the microbe has encountered. The microbe can then use this viral DNA to turn *Cas* enzymes into precision-guided weapons. The microbe copies the genetic material in each spacer into an RNA molecule. *Cas* enzymes then take up one of the RNA molecules and cradle it. Together, the viral RNA and the *Cas* enzymes drift through the cell. If they encounter genetic material from a virus that matches the CRISPR RNA, the RNA latches on tightly. The *Cas* enzymes then chop the DNA in two, preventing the virus from replicating.[19]

It turns out that many different bacteria have CRISPR systems, systems that use CRISPR in conjunction with many different "scissors" proteins, some in the Cas family and some with other obscure names, to do their work. (Cas9 has been particularly well explored.) Scientists have estimated that some kind of CRISPR system is found in 50 percent of bacterial species and 90 percent of species of archaea.[20]

During this century's first decade, Mojica and others—including scientists working for Danisco, the company that makes Dannon yogurt—explored CRISPR as a fascinating piece of natural history, a cool and previously unknown trick of Mother Nature. It was not until 2012 that CRISPR, and especially CRISPR in combination with Cas9, began to be seen as a tool for humans. (People will often write about CRISPR along with the particular associated protein they have used, so you'll see "CRISPR-Cas9," "CRISPR-Cas13," "CRISPR-cpf1," and others. For this book I will usually talk about just "CRISPR," but remember that it can only work in conjunction with one of these many associated proteins.)

It is CRISPR as a tool that is important for us, not CRISPR as an immune system. It makes it much easier, faster, and cheaper for humans to change the DNA of any living organism. This book is about the use of CRISPR to edit humans, specifically their germline genomes. But CRISPR will certainly be much more important as a tool we use to edit the rest of the biosphere; in fact, that is happening much faster with nonhumans than in humans, in large part due to our (appropriately) heightened safety concerns for editing in humans.[21]

The first publication of the idea of CRISPR as a tool came from Jennifer Doudna of the University of California, Berkeley, and Emmanuelle Charpentier, a researcher born in France who, from 2009 to 2014, was a faculty member of Umëa University in Sweden. In early June 2012, Doudna and Charpentier submitted an article to the journal *Science* on how humans could use CRISPR-Cas9 to reliably make double-stranded cuts in DNA in cells at very specific locations, determined by the guide RNA.[22] *Science* recognized the importance of the article and very quickly published it online on June 27. The last 12 words of the last sentence of the paper's abstract explains why *Science* was so interested: "Our study reveals a family of endonucleases that use dual-RNAs for site-specific DNA cleavage and *highlights the potential to exploit the system for RNA-programmable genome editing*" (emphasis added).[23] The prospect of a new tool for programmable genome editing was *very* exciting.

Often, when a scientific idea is ripe, several investigators will have similar ideas. Virginijus Šikšnys at Vilnius University in Lithuania and his group submitted an article with similar findings to the journal *Proceedings of the National Academy of Sciences* in May 2012, *before* Doudna and Charpentier.[24] *Proceedings of the National Academy of Sciences* put the paper through a peer review process that resulted in questions to the authors and substantial back and forth. That paper was ultimately published in September 2012.[25]

Science published some other important CRISPR articles in January 2013. Doudna and Charpentier had described how CRISPR with Cas9 could be a tool in bacterial (and presumably archaeal) cells. Feng Zhang and his group from the Broad Institute (a collaboration between Harvard University and the Massachusetts Institute of Technology, with the counterintuitive pronunciation of "Brōd," with a long "o"), showed that CRISPR could also work in cells from "eukaryotes," the more complicated forms of life that include amoebas, fungi, plants, and animals.[26] In the same issue, though submitted a little later, Harvard Medical School's George Church and his laboratory showed that CRISPR could be used to cut DNA in the cells of one eukaryote of particular interest to us—humans.[27]

The early uses of CRISPR-Cas9 (as the combination is called) were for cutting DNA, but scientists, those mentioned above and others, quickly began figuring out how to use CRISPR for many other purposes. This expansion of the uses of various constructs with CRISPR is still going on, with new methods, often for new purposes, announced regularly. One of the earliest, and most useful, of these extensions uses CRISPR constructs not just to cut out stretches of DNA but to replace them with other (human-chosen or human-engineered) stretches.

The irresistible (to me) analogy is to the cut-and-paste functions in word processors, particularly Microsoft Word and its "replace" function. With Word you can tell your computer to find any set of characters in a document (say "Greeley"), cut it out, and replace it with the correct version (in my case, "Greely"). In cells, the guide RNA in the CRISPR could be chosen to find a stretch of DNA that reads GTGCACCTGACTCCT-GTG. An associated protein would cut out these 24 bases, but the whole complex could include a different stretch of 24 bases, say GTGCACCTGACTCCTGAG, identical except for the next-to-last nucleotide base, which has changed from a "T" (thymine) to an "A" (adenine). Through one of several DNA repair processes, the cell will take the new DNA and put it in the place of the old DNA, thus permanently changing the cell's DNA.

This isn't a random example. That stretch of DNA is the first 24 bases of a version of the *hemoglobin-beta* gene found in millions of people. This gene provides the instructions for making a part of the hemoglobin protein in some people. People with two copies of the first version of the gene will make only abnormal hemoglobin-beta protein, will have poorly functioning hemoglobin, and will develop sickle-cell disease. The edit, putting an A in the penultimate position in place of a T, turns that into the common version of the gene, found in people who do not, and cannot, have sickle-cell disease. So CRISPR might be used to change DNA of people with sickle-cell disease to a version that would give them normal hemoglobin and hence normal blood, blood that would not cause an always painful and often life-shortening condition. (And, in fact, people are trying to do just this, including my Stanford colleague Matt Porteus, of whom you will hear more.)

Overlapping inventions, with one scientist building on the work of other scientists—"standing on the shoulders of giants"[28]—is not unusual. It is often held up as the normal, and admirable, way in which science works. The issue of multiple inventors, though, has taken on special significance here because CRISPR is so obviously important. Both fame (including prizes) and fortune (prize money but also patents and companies) are at stake. The history of, and credit for, CRISPR has become extremely controversial.

On the patent side, quite literally scores of millions of dollars have been spent on litigation over various CRISPR patents and patent applications, applications from the University of California system for Doudna and Charpentier's work and from the Broad Institute for Zhang's work.[29] The Broad filed its patent later than UC, usually the kiss of death, but it used a special "short form" kind of application that gave it priority and it got the first U.S. patent, which has forced the UC system to paddle upstream, at least in the United States. Other people, including Šikšnys, also have patent claims that may turn out to be important—or not—but thus far Zhang/the Broad versus Doudna/UC has been the main event. As I write this, Zhang and the Broad Institute are (more or less) winning in the United States while Doudna, Charpentier, and the UC system are heavily winning in Europe and China,[30] but that could change in a week, let alone in the year before these words see the light of day.

I am not a patent lawyer and try to avoid playing one on television. Predicting the outcome of patent disputes is one of the darkest of all dark arts, and I leave it to the experts, such as my former Center for Law and the Biosciences fellow, Jacob Sherkow,[31] and my colleague, the world's leading academic patent law expert, Mark Lemley. But I *am* willing to go out on a limb and say that the UC/Broad litigation is ridiculous. Truly, scores of millions, if not hundreds of millions, of dollars have been spent on this litigation. It has benefited the lawyers, many of them my former Stanford Law School students, but it is hard to see it benefiting others. Certainly, the world will little care—and little benefit from—whichever of the two gets a de jure monopoly on the use of CRISPR-Cas9 for 20 years after the patent application was filed. From a social welfare perspective, except for those with the most blind attachment to the obscure and (largely) unpredictable rules that govern who gets patents (and what those patents mean), this is just a question of who gets the rents from being able to charge higher prices for licensing CRISPR-Cas9 to firms.

Most likely, I think, is a result where whoever gets the dominant patent position will find it of little financial value. The key patents being fought over are about CRISPR-Cas9 (which is why I'm actually using "CRISPR-Cas9" here after earlier writing that I would generally avoid using it), but exciting publications have already seen the light of day about many other Cas proteins, as well as other "cutting" proteins that do not have the Cas name at all. Bacteria have been around for over 3 billion years and have evolved many immune systems with a CRISPR-like mechanism. If one institution has a patent on CRISPR-Cas9, that will only drive "invent arounds," using noninfringing cutting enzymes, or possibly even noninfringing versions of or equivalents to CRISPR. The higher the royalty or other price charged for a license, the more likely an invent-around response becomes. And the more money the licensees spend on litigating these patents, the more likely they are to need to charge high prices. It should be a classic death spiral. High prices lead customers to move to cheaper alternatives, which leads in turn to higher prices for the fewer remaining customers, which makes still more of them flee to alternatives. Madness lies along that path! (I am not an investment advisor, but I certainly wouldn't invest in any of the firms paying for the CRISPR-Cas9 litigation—even if conflict of interest concerns didn't keep me from investing in any bioscience firms.) This just seems irrational to me.

But perhaps there is some reason, albeit probably not a good one. A Nobel Prize has not yet been awarded for CRISPR (at least, as I write this—by the time you read it, that may have changed). Normally, this would not be a surprise for a relatively recent discovery. But CRISPR has been so universally adopted, and acclaimed, that people were predicting that some of its inventors would win the Nobel (either the prize in chemistry or the one in medicine or physiology) in 2017 or 2018. Some have speculated that the patent litigation has been pursued in part because Eric Lander, the director of the Broad Institute, is very eager to see his faculty member, Feng Zhang, win a part of this prize and that this may have influenced the course, and continuation, of the litigation. Lander is one of the most powerful people in genomics in the United States; the Broad Institute has long been one of the leading centers in the country for DNA sequencing. I am confident that Lander would have been a great appellate lawyer. (That's a compliment coming from me. Mainly.)

In 2016, he published an article on the history of CRISPR in the journal *Cell*, which, in the biological sciences, is esteemed at about the same very high level as *Science* and *Nature*. The article, entitled "The Heroes of CRISPR,"[32] is an excellent review of the history of CRISPR, focusing heavily on the many people who worked on the bacterial uses of CRISPR. It has, however, been roundly criticized for lauding Zhang's contributions and barely mentioning Doudna and Charpentier.[33] The first sentence of the piece, published in 2016, reads, "Three years ago, scientists reported that CRISPR technology can enable precise and efficient genome editing in living eukaryotic cells." That puts the key moment of CRISPR development in 2013, not in June 2012, and with eukaryotic cells, Zhang's work, not that of Doudna and Charpentier. The piece has many good points (including good writing—again, Lander would have been an excellent lawyer), but its treatment of Doudna and Charpentier compared with that of Zhang is, to my mind, at best, embarrassing and, at worst, shameful, particularly as the article, as published, made no mention of the conflict of interest that came from the author being Zhang's boss. (Lander was not without defenders, including the excellent science journalist Sharon Begley in an article entitled "Why Eric Lander Morphed from Science God to Punching Bag."[34])

All that being said, as a lawyer myself (though largely recovered, having last practiced more than 35 years ago), I cannot fault Lander much. And I have no inside information on whether and to what extent the Broad/UC patent fights have been motivated by a desire to enhance, through patent priorities, Zhang's position vis-à-vis Doudna and Charpentier. The Broad says that it has regularly offered to settle the dispute, through a patent pool or otherwise, but that the UC system has rebuffed its overtures.[35] Having been a general civil litigator, I know that litigants can be irrational. Perhaps it is the (always strapped for cash) UC system or the firms to which they have licensed their intellectual property who are pushing the continued trench warfare over these patents. I find the speculation that the Broad is eager to enhance Zhang's credit, fame, and chances for part of a Nobel Prize a fascinating possible explanation for the patent fight, but it is only speculation.

When will CRISPR Nobel Prizes be awarded? That is, of course, impossible to tell. The key discovery date is 2012. In recent decades, awards within five years of a discovery are uncommon but not unheard of; so are prizes 50 years after a discovery. The 2012 Nobel Prize in Medicine or Physiology provides a good example. Shinya Yamanaka won half the prize for his discovery of induced pluripotent stem cells, in mice in 2006 and in humans in 2007. He shared that prize with John Gurdon, whose work on reprogramming cells, through cloning frogs, started in the 1950s and arguably culminated in 1962. One waited six years; the other waited 50.

But there is another issue that may delay a CRISPR Nobel Prize: uncertainty about who should receive it. The science Nobel Prizes, by the organizers' rules, cannot be awarded to more than three people in any one year. And yet every scientist knows, and every nondelusional scientist truly believes, that hundreds or thousands of researchers contributed to almost every significant advance. The history of the field—whether told by Doudna and Steinberg in *A Crack in Creation*, by Lander in "The Heroes of CRISPR," or others—points out scores of people, in many countries, who contributed to the discovery. Yet only three can receive a Nobel Prize. And, as indicated in the discussion of the history of CRISPR, this raises a problem for whichever Nobel Prize group considers CRISPR. Who invented it? And who should win the Nobel Prize for CRISPR when it is limited to three people?

Personally, I think Doudna and Charpentier should be shoo-ins, for their own insights and for the work of their labs, together and separately. That seems to leave one spot. None of the talented scientists working under them in their labs are likely to win; their work will be subsumed into that of their superior's lab. That leaves at least Mojica, Šikšnys, Zhang, and Church as plausible candidates—but only one spot seems open. We shall see—or, perhaps by the time you've seen this, we will have already seen in October 2020.[36]

People have been talking about human germline genome modification for decades, though without necessarily using those words. More than 15 years ago, I reviewed two books about changing a human's DNA in ways that would get into the eggs or sperm and possibly be passed down to future generations,[1] but even then, the debate had been raging for decades. This chapter looks at those discussions up to the disclosure of He's experiment, in two parts: the early discussions of recombinant DNA technology, notably at the famous Asilomar Conference, and the more focused discussions after the development of CRISPR as a genome editing system. Along the way, it takes a look at some of the people, "CRISPR people," although not "CRISPR'd people," who were involved.

Asilomar and the Ethics of Recombinant DNA

Before the realization that DNA was the basis for human genetic inheritance and the knowledge, with Watson and Crick's discovery of DNA's structure, of the importance of DNA sequence, the discussion would not have been of "editing," as the analogy between the genome and a book (or a blueprint) did not exist. But even after Watson and Crick, and after the working out during the 1960s of how DNA in the genome "coded for" the amino acids of proteins, the discussion remained abstract. No one knew how to edit the human germline genome—or anything's genome for that matter.

That changed, a little, in 1971 with the invention of recombinant DNA. Researchers learned how to move bits of DNA from one species into another using laboratory tools. The methods were crude. Stanford Medical School biochemistry professor Paul Berg used DNA from two different species of virus, the simian virus SV40 and a bacteriophage called lambda. These viruses kept their DNA in closed-loop structures. Berg's team broke those loops in a particular location, using one of a then-recently discovered kind of enzyme called a restriction enzyme that cut DNA at specific sequences (rather like a single target version of CRISPR/CAS). Using techniques pioneered by other Stanford colleagues, the ends of these linear fragments of DNA formed from the now-open loops were made "sticky" and able to attach to other bits of DNA. These linear pieces of DNA from the two different species were mixed, and, sometimes, they combined so make a new closed loop with DNA from each species—Berg had "recombined" DNA from two different species into one piece.[2]

PAUL BERG

Berg was (and is) a biochemist's biochemist. Born in June 1926 in Brooklyn, after military service in World War II, he ended up at Penn State as an undergraduate, finishing his bachelor's degree in 1948. Largely by chance he ended up as a graduate student at Case Western Reserve University, where he got his Ph.D. in biochemistry in 1952. After postdoctoral and other fellowships in

Copenhagen, at Washington University in St. Louis, and at Cambridge University, he started his first faculty position at Washington University in 1955, working with Arthur Kornberg. In 1959 Kornberg, Berg, and several other Washington University faculty moved en masse to Stanford, where the university was moving its medical school from being mainly in San Francisco to being entirely on campus at Stanford and was opening a new biochemistry department. Paul has been a leading light at Stanford Medical School ever since.

I came to know Paul quite well starting in 1990. Stanford University celebrated the centennial of its opening with three long symposia, one in fall, one in winter, and one in spring quarter. The winter quarter symposium was to be on the then-new Human Genome Project. The symposium planning committee was chaired by Berg, joined by David Botstein, then chair of Stanford genetics department, and Lucy Shapiro, the Founding Chair of Stanford's then-new department of developmental biology. They wanted a law professor on the committee with them. One of my colleagues suggested me, probably, I've always joked, because I was the only Stanford law professor at the time who could spell DNA (let alone deoxyribonucleic acid).

I accepted and had a wonderful time on the committee. It was my first real close-up look at eminent scientists. These folks were not only brilliant, but enjoyable. I soaked up enormous amounts of information every time we met, I gained three friends who would patiently explain scientific issues to me, and I had great fun. I also ended up being chosen to give a talk at the symposium, held January 11 through 13, 1991. I gave a talk on the likely effects of genetic information on the U.S. health care financing system, the first time I had given a talk about genetics. (It turned into my first publication on genetics, a book chapter.[3]) I met amazing people through the conference, people whose research I read about for years and people whose interactions with me would change my life. And it cemented my relationship with Paul Berg, a crucial mentor for me.

Berg did not immediately take the next step of then trying to move such recombinant DNA into a living organism. That was done first by two other Bay Area scientists, UC San Francisco (UCSF) professor Herb Boyer and Berg's Stanford Medical School colleague Stan Cohen of the school's genetics department. By spring 1973 they had moved an antibiotic resistance gene into *E. coli* bacteria. Soon they moved DNA from a staphylococcus bacteria into *E. coli* and eventually did the same with DNA from a frog species.[4]

There has long been some dispute over who deserves how much credit for the invention of recombinant DNA.[5] As with CRISPR, many researchers contributed. Some have argued that others were as or even more deserving of a recognition for inventing recombinant DNA as Berg, notably Cohen and Boyer or Berg's fellow Stanford biochemistry department researchers, Janet Mertz and Ron Davis. This debate has, over the years, had some special bite at Stanford Medical School. Berg was a member of the School's biochemistry department, and Cohen, a professor of the genetics department. Although it is fading with time, I remember hearing (*mainly*, but not entirely, good-natured) ribbing between the departments over which should get credit.

As it happened, in October 1980 Berg won a share of the Nobel Prize for Chemistry for inventing recombinant DNA. Berg shared the prize with Frederick Sanger and Walter (Wally) Gilbert, who were awarded half the prize money for inventing techniques to determine the sequence of DNA. As with CRISPR today, the Nobel Committee allows no more than three people to be awarded a prize in most of its categories. Once the chemistry committee decided to recognize Sanger and Gilbert for sequencing—each had made substantial progress in very different ways—that left only a slot for one inventor of recombinant

DNA if that accomplishment were to be awarded the prize that year.

Ironically, exactly seven weeks later, the U.S. Patent and Trademark Office granted U.S. Patent No. 4,237,224, "Process for producing biologically functional molecular chimeras," to two inventors, Cohen and Boyer. Their interests in the patent were assigned to Stanford and the UC system, respectively. Stanford managed the patent for the two universities. During its 17-year lifetime, it brought the two universities over $250 million.[6] (More than a quarter of Stanford's royalties went to the genetics department with an equal amount to the medical school—none went directly to the biochemistry department.) Both Cohen and Boyer received some patent royalties through their universities; Boyer cofounded Genentech in part based on a license of that patent.

You may well be wondering what this digression into the history of recombinant DNA, patenting, and Nobel Prizes—as interesting as its parallels to CRISPR may be—has to do with ethics, the topic of this chapter. As far as Berg's Nobel Prize goes, no one doubts that he and his lab made major contributions to the field and were driving forces in its advance, but Berg had another role that made him stand out from the rest of the recombinant DNA crowd. He was a leader, arguably *the* leader, in organizing a temporary moratorium on recombinant DNA research and in organizing and running the famous 1975 Asilomar Conference on recombinant DNA at which the moratorium was discussed. And the Asilomar Conference is an essential part of this story.

The Asilomar Conference, or, to give it its full name, the Asilomar Conference on Recombinant DNA Molecules, was held on February 24, 25, and 26, 1975, at the Asilomar State Beach and Conference Grounds, an unusual unit of the California State

Park system, located on the coast just south of Monterey, California (and one of the loveliest places in the world).[7] It had been spawned in June 1973 at a Gordon Conference on the topic of nucleic acids.

Gordon Conferences are prestigious scientific meetings on frontier research in biology, chemistry, and the physical sciences. The first Gordon Conference was held in 1931, growing out of summer conferences held by the chemistry department at Johns Hopkins University in the 1920s. Initiated by Johns Hopkins professor Neil Gordon, the nonprofit organization that now runs them holds more than 300 conferences or seminars a year, usually in scenic and isolated locations.[8] At these meetings, small invited groups of researchers come together "off the record" to discuss their field. (To my regret, I have never gone to one.)

From June 11 to 15, 1973, a Gordon Conference on Nucleic Acids was held in New Hampton, New Hampshire. Talk turned to concerns about the potential safety hazards of recombinant molecules. A majority of those attending the conference voted to send a letter expressing their concerns to both the president of the U.S. National Academy of Sciences and the president of the National Institute of Medicine (now the National Academy of Medicine), as well as to publicize their letter more widely. The letter was also published in the September 21, 1973, issue of *Science*.[9] (No earlier online publication in those days.) The letter read, in part:

> Certain such hybrid molecules may prove hazardous to laboratory workers and to the public. Although no hazard has yet been established, prudence suggests that the potential hazard be seriously considered. A majority of those attending the Conference voted to communicate their concern in this matter to you and to the President of the Institute of Medicine (to whom this letter is also being sent). The conferees suggested that the Academies establish a study

committee to consider this problem and to recommend specific actions or guidelines, should that seem appropriate.

The leadership of the National Academy of Sciences and the National Institute of Medicine took the letter seriously and appointed such a committee. That committee published its own letter in *Science* on July 26, 1974.[10] This letter read, in relevant part:

> The undersigned members of a committee, acting on behalf of and with the endorsement of the Assembly of Life Sciences of the National Research Council on this matter, propose the following recommendations.
>
> First, and most important that until the potential hazards of such recombinant DNA molecule have been better evaluated or until adequate methods are developed for preventing their spread, scientists throughout the world join with the members of this committee in voluntarily deferring the following types of [recombinant DNA] experiments . . .

Additional recommendations advised caution with respect to some other types of recombinant experiments, asked the National Institutes of Health (NIH) to consider establishing an advisory committee on the topic, and said that "an international meeting of involved scientists from all over the world should be convened early in the coming year to review scientific progress in this area and to further discuss appropriate ways to deal with the potential biohazards of recombinant DNA molecules." The letter was signed by the 11 members of the committee, with Paul Berg listed first, as chair. Among the others were Herb Boyer, Stan Cohen, and Ron Davis from the Bay Area's recombinant DNA laboratories, as well as Jim Watson, codiscoverer of the structure of DNA, and a young biologist named David Baltimore, one of the discoverers of how some cells can turn RNA into DNA (for which he won a Nobel Prize).

DAVID BALTIMORE

David Baltimore was born in March 1938 in New York City and raised in Queens until second grade, when his parents moved to the suburbs in search of better schools. He received his bachelor's degree from Swarthmore College in 1960. He started his graduate work at the Massachusetts Institute of Technology but moved to Rockefeller University, where he received his Ph.D. in 1964 for work with viruses. He went back to MIT for a postdoctoral fellowship, then got training at Albert Einstein College of Medicine before moving, in 1965, to a postdoc at the then-new Salk Institute in San Diego. By 1968 he was back at MIT, this time on the faculty, and there, in 1970, at the age of 32, he discovered reverse transcriptase. This enzyme disproves, or at least qualifies, the "Central Dogma" of molecular biology, announced by the venerable Francis Crick: DNA makes RNA makes protein. Instead, reverse transcriptase uses RNA to make DNA. For this discovery, Baltimore and Howard Temin (who had found the same thing independently) shared part of the Nobel Prize in Medicine or Physiology in 1975. Baltimore was 37.

Baltimore stayed at MIT until 1982 when he helped found the Whitehead Institute for Biomedical Research, an independent research center closely connected to MIT. He remained there, while being simultaneously a professor at MIT, until 1990, when he became president of the Rockefeller University in Manhattan, but only for 18 months. He resigned the presidency of Rockefeller in the wake of a research scandal involving a coauthor from MIT.[11] He returned to MIT until 1997 when he was named president of the California Institute of Technology. He served until 2006, when he resigned as president in the wake of research fraud committed by one of his postdocs. He has remained on the Caltech faculty since then. Over the years, Baltimore has taken part in various policy issues, from Asilomar to human germline genome editing, as well as cochairing the National Academies' Committee on Science, Technology, and Law (CSTL).

> Baltimore is a short man who, even in his early 80s, radiates power and energy. He is a force of nature, and, as I think I'll put it, he is not afraid to lead. I first began to get to know him well in 2013 and 2014 through the CSTL. I must admit that I did not immediately take to him (when pushed, like a donkey, I tend to sit down), but with more experience with him I have seen and appreciated his less public and more contemplative side. And I have come to quite like and respect him.

The National Academy of Sciences appointed a five person committee to organize this international meeting. Paul Berg was the chair, joined by David Baltimore, Sydney Brenner, Richard Roblin, and Maxine Singer, all eminent scientists. The result was the Asilomar Conference.

What *was* the Asilomar Conference? It was the best of things, and it was the worst of things. Almost from before it ended, it was lauded as a wonderful example of scientific responsibility and self-governance and denounced as a terrible example of scientific hubris and self-interest. It has been the subject of histories, revisionist histories, and rerevisionist histories. It has been the model for other similar meetings, at Asilomar and elsewhere, though none of them achieved its fame—or infamy. It has, in short, been everything but forgotten.

But, to be concrete, it actually was a three-day conference with about 140 attendees from around the world. The attendees were mainly scientists doing, or planning, research in recombinant DNA but there were several government officials, 12 journalists, and four lawyers.[12]

The meeting was held at the Asilomar Conference Grounds, located on nine acres of land at the westernmost tip of the Monterey Peninsula, between the town of Pacific Grove to the north and what is now the Pebble Beach development to the south. The grounds were built starting in 1913 as a home for the annual conference of the Young Women's Christian Association of the Pacific Coast and were designed by America's first major female architect, Julia Hunt Morgan (later the main architect for William Randolph Hearst's palatial estate, known as Hearst Castle, farther south along that coast).[13] She designed 16 buildings for the property, largely in the Arts and Crafts or California Craftsman style, 13 of which still stand. The property was never financially self-supporting for the YWCA, which, during the Depression, closed it and tried to sell it. It went through various empty periods, interspersed with different attempts to manage it profitably before, in 1956, it was purchased by the California State Park system. It was joined to state park–owned beach, tide pool, and dune property adjacent to it to become the Asilomar State Beach and Conference Grounds. The conference center was declared a National Historic Landmark in 1987 (although not for hosting the Asilomar Conference).[14]

For three days in February 1975 the scientists made presentations on the science, its risks, their significance, and what could be done to limit them. On the second evening some of the lawyers made an eye-opening (to the scientists) presentation on the possible tort liability of the experimenters should things go wrong. That night, the organizing committee worked late writing up a document with recommendations for how to handle these experiments. Precautions were called for, and different levels of risk and types of experiments were identified, including some that the document said should not be done, at least at that time. The recommendations were debated and adopted by a majority vote of the attendees on the meeting's last day. A more complete summary report of the conference and its recommendations was published in the *Proceedings of the National Academy of Sciences* four months later.[15]

But the news didn't wait for that scientific publication. The 12 invited journalists had promised not to report until the conference ended, but then they reported a lot. Their publications, and others, carried stories about the conference, in publications as different as the *New York Times, Wall Street Journal, Frankfurter Allgemeine Zeitung, Nature*, and *Rolling Stone*.[16] And the world was talking about recombinant DNA and about science's effort to police itself.

Asilomar does have some ironies. First, it focused on the physical safety risks of recombinant DNA research, the chance that a life-form with recombinant DNA could harm lab workers, the public, or the environment. It did not discuss broader questions—of playing God, designer babies, hubris—that, then and now, would fascinate, and scare, the public.

Second, this exercise in scientific self-regulation actually became enforceable (in the United States) by a government action almost immediately after the conference ended. Its recommendations were adopted by the NIH, acting through a Recombinant DNA Advisory Committee (RAC) that Donald Frederickson, then NIH director, had appointed shortly before the Conference. The RAC has had a long and shifting career, surviving, in spite of increasing government efforts to eliminate it over the last several decades, at least until 2019.[17]

And yet the idea that this was self-regulation by Science was not wrong. Asilomar helped lead to the rejection of legislative efforts to restrict or stop recombinant DNA research, in the Cambridge, Massachusetts, city council and in the U.S. Senate. Regulation was left to the NIH, an organization dominated by scientists who would at least understand the researchers' perspectives.

The CRISPR Discussions

Now we come to back to human germline genome editing. At 11:45 a.m. on October 2, 2014, I received an email from Jennifer Doudna. Doudna was already well on her way to being a very famous person. I had known her name since 2012 and had been following, to some extent, CRISPR's progress, but I had never met or, as far as I know, been in the same room with her. She wrote, on behalf of an organizing group, as follows:

> We (Mike Botchan, Jennifer Doudna, Jacob Corn, Ed Penhoet and Jonathan Weissman) are hosting a one-day workshop at the Carneros Inn in Napa Valley to discuss the bioethical issues raised by the explosion in new genomic editing methods. The purpose of this meeting is to ascertain which critical issues will warrant particular attention as the field expands into ethical territories and scientific fields that were first raised decades ago when the cloning revolution began. We would be delighted if you can join us for a day of lively discussion and brainstorming.

Twenty-three minutes later, I wrote back and said "yes" (with some questions about a scheduled flight back to the Bay Area and an upcoming, but then unscheduled, hip replacement surgery). I was enthusiastic: "Barring those two contingencies (as well as meteorites, World War III, the big earthquake, and all the other contingencies), *yes.*"

Why was I invited? I didn't know Doudna or Botchan, Corn, or Weissman—I think I'd met Penhoet once or twice. Maybe they knew me by reputation, although I hadn't written anything at that point about CRISPR or, in any detail, gene editing. But I did know one of her invitees, Paul Berg, I suspect he recommended me to Doudna, probably with support from my fellow law professor, Alta Charo.

None of the bad contingencies came to pass, and so January 24 found me, on a bright, clear, warm winter's day, at a resort called the Carneros Inn in southern Napa Valley. Doudna had convened the event through the UC Berkeley Innovative Genomics Institute, which she directed (and directs). It was a meeting of a small working group. The 14 principals included mainly noted scientists, such as Nobel Prize winners Paul Berg and David Baltimore, and the soon-to-be-named dean of Harvard Medical School, stem cell researcher George Daley. Other scientist-participants included Doudna's fellow organizers, Botchan, Corn, Penhoet, and Weissman, and Betsy Nabel from Harvard, Jennifer Puck from UCSF, and Keith Yamamoto from UCSF, plus one video journalist, John Rubin from Tangled Bank Studios. (Kathy Hudson, Deputy Director for Science, Outreach and Policy at NIH, was supposed to attend but had to pull out at the last minute.) My friend Alta Charo of the University of Wisconsin and I were the only two lawyers/bioethicists at the meeting. Several students, postdocs, or staffers were also present and may have taken part in some of the substantive conversations.

The presence of Baltimore and Berg was powerfully evocative for me; they had been two of the five people on the organizing committee for the Asilomar meeting. As I pointed out at the time, Asilomar had taken place in late February 1975, 40 years, less a single month, earlier.[18]

I took moderately detailed notes on my laptop, a habit of mine at meetings. These totaled about 900 words for the eight-hour meeting. I cannot, however, make maximal use of them here as the meeting was held under a version of Chatham House Rules—the substance of what was said can be discussed outside the group but statements cannot be attributed to individuals. I will abide by those rules (albeit with some regrets).

The workshop started with a welcome from Doudna and introductions, then a workshop charge and goal from Doudna and Botchan. Paul Berg and David Baltimore led a spirited discussion of the lessons of Asilomar (from the horses' mouths, the agenda should have said). George Daley gave a wide-ranging talk on the future of somatic cell gene therapy, stem cell research and treatments, cloning, and in vitro gametogenesis, among other things. Then Alta Charo and I split up the legal aspects of genome engineering; she took the nonhuman aspects and I took the human ones. After a lunch break, Jacob Corn and Betsy Nabel led a discussion of emerging scenarios.

This description, which follows the premeeting agenda, makes the event sound like presentations in a lecture hall. But it wasn't. The room was relatively small and held a long table just big enough for the 14 people at it, each of whom had things to say. These presentations were punctuated by frequent questions, pronouncements, and debate. Berg and Baltimore, in particular, gave their opinions freely and often. In my notes, I called them the "old bulls," but, as they were the two grand old men at the table—and both very smart and experienced—no one complained. By 3:00 the agenda called for the group to start to "draft a white paper position statement."

Two and only two of the participants—whom I am not allowed to name—argued for at least some attention to the nonhuman uses of CRISPR, which they thought would, at least in the next few decades, likely be much more important than the human uses. The tide, however, ran strongly the other way, not just in favor of focusing on human uses but specifically on human germline uses. Some, whom I am also not able to name, expressed the view that, since Asilomar and the recombinant

DNA debates, Science had promised that the human germline would not be manipulated and that this was a crucial issue to confront. And it is the case that, although they were not discussed in any detail at the meeting, at least not openly, rumors abounded that Chinese scientists were about to announce that they used CRISPR to modify human embryos. Those rumors, whatever we thought of them, added a certain urgency to the meeting.

The group reached consensus surprisingly quickly, dividing its attention into three main areas: in vitro research, somatic cell therapy, and germline genome editing. We all agreed that CRISPR had an important and immediate role to play with regard to in vitro research in humans, had great promise for use to edit the genes of people born with genetic diseases, but should not be used—at least for the time being—for germline editing. We broke up just about on time—maybe early though my notes don't give the time—with the goal that some of the participants would turn that consensus into a paper. Then we all, the participants and in some cases their partners or guests (my wife came), had a delightful dinner in downtown Napa, with the participants, I believe, all feeling we had put in a good day's work. (I certainly did.) We went back to the Inn, slept at the resort, and dispersed the next day.

JENNIFER DOUDNA

That was the first time I met Jennifer Doudna,[19] whom I have been fortunate to interact with another dozen or so times since then. Jennifer was born in Washington, DC, in 1964, but from the age of seven, she grew up in Hilo, on the northeast (rainy) corner of the Big Island of Hawai'i where her father was a professor of American literature at the University of Hawai'i, Hilo. She graduated with a bachelor's degree in biochemistry from Pomona College (home of the Sagehens) in 1985. (Like Baltimore, she is a graduate of a liberal arts college and not a research university.) She received her Ph.D. in biological chemistry and molecular pharmacology in 1989 from Harvard Medical School, writing a dissertation on RNA under the supervision of Jack Szostak, who won a Nobel Prize in 2009. After two years of postdocs in Boston, she spent four years, until 1994, as a postdoc at the University of Colorado, working with Thomas Cech, who had already won a Nobel Prize in 1989. Her first appointment as a professor was at Yale, but she moved to UC Berkeley in 2002.

Doudna did basic science research on RNA, *not* a sexy subject. Her fame dates from her June 2012 paper on CRISPR with Emmanuelle Charpentier, but she had already been recognized as an eminent scientist. She was elected to the National Academy of Sciences in 2002, at age 38, to the American Academy of Arts and Sciences in 2003, and to the Institute of Medicine (now the National Academy of Medicine) in 2010, all *before* her pathbreaking work on CRISPR.

Doudna is a tall woman with light colored hair. She has no bluster; she speaks quietly and calmly but commands respect with what she says and how she says it. She is thoughtful and smiles a lot. I liked her immediately and, after all our interactions since January 2015, have only come to like her more. She seems totally unspoiled by fame. (I've also admired how she has become an excellent public speaker and "science explainer" with those several years of practice.) It is probably worth noting, though, that between prizes for CRISPR, interests in companies working on CRISPR, speaking fees, potential royalties from the UC system for CRISPR patents, and other income—including the royalties on her coauthored book, *A Crack in Creation*—she has undoubtedly made a lot of money as a result of CRISPR, with interests in the technology that are likely to continue to pay off.

Over the next several weeks, the Napa meeting participants would get regular updates, with new drafts of a publication to comment upon. I was not one of the drafters but, along with many of the participants, offered substantive and stylistic advice to those preparing the publication. And, remarkably, in fewer than eight weeks after the Napa meeting, those drafts, with the addition of a few more authors, became an article published online in *Science* on March 19, 2015.[20] (The authors were listed in alphabetical order, but it did feel right to me that the alphabet made the citation start with the two Asilomar veterans, David Baltimore and Paul Berg.)

That article made four recommendations:

Strongly discourage, even in those countries with lax jurisdictions where it might be permitted, any attempts at germline genome modification for clinical application in humans, while societal, environmental, and ethical implications of such activity are discussed among scientific and governmental organizations . . .

Create forums in which experts from the scientific and bioethics communities can provide information and education about this new era of human biology, the issues accompanying the risks and rewards of using such powerful technology for a wide variety of applications including the potential to treat or cure human genetic disease, and the attendant ethical, social, and legal implications of genome modification.

Encourage and support transparent research to evaluate the efficacy and specificity of CRISPR-Cas9 genome engineering technology in human and nonhuman model systems relevant to its potential applications for germline gene therapy . . .

Convene a globally representative group of developers and users of genome engineering technology and experts in genetics, law, and bioethics, as well as members of the scientific community, the public, and relevant government agencies and interest groups, to further consider these important issues, and where appropriate, recommend policies.

In the meantime, another article on the use of CRISPR in humans had appeared online the week before in *Nature*, calling for an absolute ban on germline modifications in humans, in part expressly to prevent germline genome editing from souring the public on the good uses of CRISPR in somatic cell gene therapy.[21] And, on April 18, the shoe dropped with the publication of an article by Chinese scientists, discussed in chapter 1, reporting that they had, with some limited success, used CRISPR to edit (nonviable) human embryos.[22]

This narrative should sound familiar from the discussion of Asilomar. A small group of leading researchers meets at a workshop and worries about the ethical issues raised by a new technology. They publish their concerns and call for, among other things, more discussion. And the next step unfolded in the same general way as the Asilomar Conference process: the U.S. National Academy of Sciences and National Academy of Medicine got involved. On May 18, 2015, the presidents of the two academies announced the creation of a Human Genome Editing Initiative along with a conference to be held at a yet to be determined date and place.[23] (At least one newspaper article says, without a stated source, that Doudna wanted the conference to be organized by the Howard Hughes Medical Institute, a large science philanthropy, but Baltimore had pushed for National Academies.[24]) On June 15, the Academies announced the membership of a 14-person advisory group for this initiative. It included Baltimore, Berg, Charo, Doudna, Penhoet, and Yamamoto from the Napa meeting.[25]

The details of the initiative's conference were announced on September 14—an International Summit on Human Gene Editing, to be held in Washington, DC, from December 1 to 3, 2015.[26] This was jointly sponsored by the National Academy of Sciences, the National Academy of Medicine, the Royal Society of the United Kingdom, and the Chinese Academy of Sciences. The organizing committee, announced the same day, included David Baltimore as chair, along with Napa meeting alumni Paul Berg, George Daley, and Jennifer Doudna. Other notable additions included Eric Lander from the Broad Institute, Canadian bioethicist Françoise Baylis, and British researcher Robin Lovell-Badge.

This highly publicized two-and-a-half-day event was held in the main auditorium of the Academies' impressive 1924 building, lodged between the National Mall to its south and the State Department to its north. It included scores of speakers and panelists and even more journalists. (Information about this first summit, including the agenda and links to video recordings of the entire event, can be found on the initiative's website,[27] and the Academies' summary of the summit, prepared by science journalist Steve Olson, is also available.[28])

My own role (other than engaged listening) was limited to moderating the very last panel of the event, on governance, regulation, and control. That panel was followed by lunch and "closing thoughts" from organizing committee chair David Baltimore. Baltimore's closing thoughts included a statement from its organizing committee, prepared, no doubt, over the previous day and night.[29] This statement was *not* an official position of the National Academy of Sciences, the National Academy of Medicine, or other sponsors, as those entities were quick to stress, but was the organizing committee speaking for itself. The statement's recommendations were quite similar to those of the March 2015 *Science* article—perhaps not surprisingly, since David Baltimore chaired the organizing committee.

The short statement, under 1,000 words, encouraged basic and clinical research as well as somatic cell clinical uses. As to germline uses, the committee concluded,

> It would be irresponsible to proceed with any clinical use of germline editing unless and until (i) the relevant safety and efficacy issues have been resolved, based on appropriate understanding and balancing of risks, potential benefits, and alternatives, and (ii) there is broad societal consensus about the appropriateness of the proposed application. Moreover, any clinical use should proceed only under appropriate regulatory oversight. At present, these criteria have not been met for any proposed clinical use: the safety issues have not yet been adequately explored; the cases of most compelling benefit are limited; and many nations have legislative or regulatory bans on germline modification. However, as scientific knowledge advances and societal views evolve, the clinical use of germline editing should be revisited on a regular basis.

It then pointed out the international importance of the question:

> While each nation ultimately has the authority to regulate activities under its jurisdiction, the human genome is shared among all nations. The international community should strive to establish norms concerning acceptable uses of human germline editing and to harmonize regulations, in order to discourage unacceptable activities while advancing human health and welfare.

And it called for an ongoing international forum to continue discussing these issues, with a "wide range of perspectives and expertise—including from biomedical scientists, social scientists, ethicists, health care providers, patients and their families, people with disabilities, policymakers, regulators, research funders, faith leaders, public interest advocates, industry representatives, and members of the general public."

ALTA CHARO

I can't quite figure out how long I've known R. Alta Charo (don't ask about the "R"; she doesn't use it); it has been at least since 2003 but probably earlier. Charo is a law (and bioethics) professor at the University of Wisconsin at Madison. She grew up in New York City and graduated from Harvard in 1979 with a biology degree, then got her JD from Columbia in 1982. At some point during those years, she spent at least some time doing laboratory work looking at chromosome numbers in embryos. This was an omen of a career spent in bioethics and particularly on reproductive and stem cell issues. After law school she worked, among other places, in the Biological Applications Program of the late (as of 1995) and lamented (still today, by some of us) Congressional Office of Technology Assessment. She joined the Wisconsin faculty in 1989, and, with the exception of various visits and leaves, she has been there ever since.

In spite of her Madison, Wisconsin, address, much of Charo's work has been conducted in Washington, DC. She served on an NIH Human Embryo Research Panel in the early 1990s and on President Clinton's National Bioethics Advisory Commission. She has served on innumerable National Academy of Sciences or Medicine committees and boards; she was elected a member of the National Academy of Medicine in 2006. She worked on President Obama's transition team for the U.S. Department of Health and Human Services (HHS) and stayed in Washington until 2011, serving as a senior policy advisor on emerging technology issues at the U.S. Food and Drug Administration (FDA) before returning to Wisconsin.

Charo is regularly asked to do these things mainly because she is smart and hardworking, with good writing and political skills, but I suspect also because she is fun to have around. She is verbally quick, with a flashing sense of humor, sometimes self-deprecating (as in the jokes about how short she is). She loves various iterations of *Star Trek* and composing satirical lyrics to show tunes. (I live in fear that she has one with my name in it.)

The National Academy of Sciences and the National Academy of Medicine followed up the summit with a February 14, 2017, consensus report, written by a 22-person committee, which Alta Charo cochaired with MIT scientist Richard Hynes.

This Valentine's Day report, like most Academies reports, was the result of many public and private meetings and a long engagement with many peer reviews.[30] The report, as do all such National Academies' reports, states the "official position" of the Academies. The report had chapters on basic research, somatic genome editing, heritable (germline) genome editing, enhancement, and public engagement. It offered clear, and strong, conclusions on germline genome editing: "In particular, clinical trials using heritable germline editing should be permitted only if done within a regulatory framework that includes the following [10] criteria and structures. . . ." These criteria deal at some length with the conditions to be addressed by the editing and the mechanisms by which it should be considered and carried out:

1. absence of reasonable alternatives;
2. restriction to preventing a serious disease or condition;
3. restriction to editing genes that have been convincingly demonstrated to cause or to strongly predispose to the disease or condition;
4. restriction to converting such genes to versions that are prevalent in the population and are known to be associated with ordinary health with little or no evidence of adverse effects;
5. availability of credible preclinical and/or clinical data on risks and potential health benefits of the procedures;
6. ongoing, rigorous oversight during clinical trials of the effects of the procedure on the health and safety of the research participants;
7. comprehensive plans for long-term, multigenerational follow-up while still respecting personal autonomy;

8. maximum transparency consistent with patient privacy;

9. continued reassessment of both health and societal benefits and risks, with broad ongoing participation and input by the public; and

10. reliable oversight mechanisms to prevent extension to uses other than preventing a serious disease or condition.[31]

The report noted the difficulties of defining "enhancement":

For example, using genome editing to lower the cholesterol level of someone with abnormally high cholesterol might be considered prevention of heart disease, but using it to lower cholesterol that is in the desirable range is less easily characterized, and would either intervention differ from the current use of statins?[32]

That chapter concluded that "genome editing for purposes other than treatment or prevention of disease and disability should not proceed at this time, and that it is essential for these public discussions to precede any decisions about whether, or how, to pursue clinical trials of such applications."[33]

The authors pushed strongly for public engagement in decisions about genome editing, specifically recommending that, "[w]ith respect to heritable germline editing, broad participation and input by the public and ongoing reassessment of both health and societal benefits and risks are particularly critical conditions for approval of clinical trials."

Some commentaries on the National Academy of Sciences/ National Academy of Medicine report saw it as opposing heritable genome editing; others saw it as permissive.[34] Meanwhile, in the United Kingdom, the Nuffield Council, an independent nonprofit bioethics advisory group, issued two relevant reports. In 2016, it issued a general report entitled *Genome Editing: An Ethical Review*.[35] And then, in July 2018, it issued *Genome Editing and Human Reproduction*, saying, "We conclude that the potential use of heritable genome editing interventions to influence the characteristics of future generations could be ethically acceptable."[36] Their eight requirements and recommendations overlap some with those of the Valentine's Day report but, interestingly, include more explicit discussion of social issues and of specific regulatory recommendations (albeit for the United Kingdom only). It wrote,

> We conclude that the potential use of heritable genome editing interventions to influence the characteristics of future generations could be ethically acceptable in some circumstances, so long as:
>
> - it is intended to secure, and is consistent with, the welfare of a person who may be born as a consequence of interventions using genome edited cells; and
>
> - it is consistent with social justice and solidarity, i.e. it should not be expected to increase disadvantage, discrimination, or division in society.
>
> - We recommend that research should be carried out on the safety and feasibility of heritable genome editing interventions to establish standards for clinical use.
>
> - We recommend that social research should be carried out to develop greater understanding of the implications of genome editing for the welfare of the future person.
>
> - We recommend that before any move is made to amend UK legislation to permit heritable genome editing interventions, there should be sufficient opportunity for broad and inclusive societal debate.
>
> - We recommend the establishment of an independent UK body to promote public debate on the use of genomic and related technologies to respond to societal challenges; to help to identify and understand the public interests at stake; and to monitor social, cultural, legal, and health impacts.
>
> - We recommend that governments in the UK and elsewhere should work with international human rights institutions, such

as the Council of Europe and UNESCO [United Nations Educational, Scientific, and Cultural Organization], to promote international dialogue and to develop a framework for international governance of heritable genome editing interventions.

- We recommend that heritable genome editing interventions should only be licensed on a case-by-case basis subject to: assessment of the risks of adverse clinical outcomes for the future person by a national competent authority (in the UK, the HFEA [Human Fertilisation and Embryology Authority]); and strict regulation and oversight, including long-term monitoring of the effects on individuals and social impacts.

It's worth noting that three of those eight points call for public debate and discussion or international dialogue.

The next major event in the ethical assessment of human germline genome editing was part of the efforts to have a global conversation about the topic. The Second International Summit on Human Genome Editing was scheduled to take place in Hong Kong from November 27 through 29, 2018, with an (albeit nonexclusive) emphasis on speakers from Asian countries. The National Academy of Sciences, the National Academy of Medicine, and the Royal Society of the United Kingdom once again were three of the sponsors. For this event, however, the initial fourth sponsor, the Chinese Academy of Sciences, pulled out, for reasons unclear (at least to me), about a year before the event and was replaced by the Academy of Sciences of Hong Kong.[37] After one more background chapter, on the legal setting for human germline genome editing before that meeting, we will go to Hong Kong for that Second International Summit.

5   The Law of CRISPR'd Babies before He

In chapter 4, I quoted the recommendations from the National Academies report and the Nuffield report at length because they are the most serious, searching, and detailed assessments of human germline genome editing from before the He Jiankui affair. Between them, they contain 18 thoughtful, careful guidelines. Neither report, however, has the force of law—in the United States, in the United Kingdom, or anywhere else. So— what does the law say?

As always, the answer is complicated. At least three different kinds of law are relevant, each one confronting human germline genome editing with a different level of specificity and with each one usually varying to some extent by country. We have to think about the laws of human subjects research, the laws of development and approval of medical treatments, and finally of laws directly about human germline genome editing. Think of them as similar to laws broadly against hurting someone, laws more narrowly against assaulting someone with a deadly weapon, and laws specifically against shooting someone with an automatic rifle.

The Law of Human Subjects Research

Most countries that participate substantially in human subjects research have similar laws and regulations about it. Part of this is because of international codes from private organizations. The most important is probably the 1964 Helsinki Declaration of the World Medical Association (WMA), along with its various amendments in different meeting locations over the decades,[1] and the "International Ethical Guidelines for Health-related Research Involving Humans" of the Council for International Organizations of Medical Sciences (CIOMS) in collaboration with the World Health Organization (WHO).[2] In addition, in 2006 UNESCO adopted a "Universal Declaration on Bioethics and Human Rights," which includes, but is not limited to, ethical principles for human subjects research.[3]

Those codes owe a debt to an actual legal document, although an unusual one: the "Nuremburg Code," part of an August 19, 1947, decision by the American Military Tribunal in the American-Occupied Zone of Germany in the case of *U.S. v. Karl Brandt*, usually known as "the Doctors' Trial."[4] This case tried 23 defendants (20 of whom were physicians) for their "experiments" with euthanasia and other mistreatment of prisoners of war, citizens of occupied countries, and citizens of Germany. They were charged with conspiracy to commit war crimes and crimes against humanity, war crimes, crimes against humanity, and membership in a criminal organization (the SS or Schutz-staffel). Seven of the defendants were acquitted of all counts while 16 were found guilty on one or more charges. Of those found guilty, seven were hanged by the neck until dead.

Some of the defendants claimed that their actions were not importantly different from medical research in other times and countries. In apparent response, the verdict included a section entitled "Permissible Medical Experiments."[5] Its 10 points, expressed in just over 500 words, became known as "the Nuremburg Code." They included such foundational principles as voluntary consent, an absence of coercion, properly formulated research, a proper balance of risk and potential benefit, and a participants' right to withdraw from the research. This "Code" was not "law" in a particularly deep way—it was not adopted by any legislature—but was the conclusion of a military court in a jurisdiction that, at this point, hasn't existed for over 65 years. Nonetheless, it has been important, as, with various modifications, its basic principles have been incorporated in subsequent codes.

The Helsinki Declaration was adopted in 1964 by the WMA, a group of various national medical associations, such as the American Medical Association. Its strictures are addressed primarily to physicians, but the WMA encourages their adoption by others involved in medical research. It has been amended nine times since then at WMA General Assemblies, though mainly in its details. Its 37 principles require about 2,200 words.

The CIOMS Guidelines were first adopted in 1982; their fourth version was adopted in 2016. Their 25 guidelines, with commentary, take up about 100 pages. (As is clear, an inflationary force has been at work on these codes, though, to be fair, with experience has come more appreciation for nuance. Of course, more authors and editors can also yield more words.) Written with the WHO, these guidelines say, "The ethical principles set forth in these Guidelines should be upheld in the ethical review of research protocols. The ethical principles are regarded as universal."[6]

Apart from these international guidelines, another force has pushed toward similar rules. In 1990 the United States, the

European Union, and Japan joined together in the International Council for Harmonisation of Technical Requirements for Registration of Pharmaceuticals for Human Use, renamed the International Council for Harmonisation of Technical Requirements for Pharmaceuticals for Human Use (ICH) in 2015 when it became a Swiss nonprofit organization.[7] Its goal has been to harmonize requirements so that clinical trials in one jurisdiction will be acceptable in the others.

The ICH looks at both scientific and ethical standards in its provisions on Good Clinical Practices (the ICH-GCP). Its standards are part of the formal regulatory system in the European Union, acting through the European Medicines Agency, and, although they are not "law" in the United States, they parallel American legal requirements and will be adhered to in American clinical trials whose sponsors want approval in the European Union, Japan, or other ICH member countries. Clinical trials conducted in places outside the United States, European Union, and Japan will often follow the ICH-GCP in order to maximize the trials' international value. Today the three founding members have been joined by regulatory agencies in Canada, Switzerland, Brazil, Singapore, South Korea, Taiwan, and, since 2016, China.[8]

The international codes and the ICH all push countries to systems of human subjects protection that demand informed consent, require a balancing of risks and benefits, and demand oversight and advance approval of human subjects research by some kind of ethics committee. In the United States this system is implemented through the "Common Rule," a regulation adopted in common by nearly 20 departments and federal agencies. This sets out detailed requirements for human subjects research.[9] The reviewing ethics committees in the United

States are known as "institutional review boards" (IRBs) and are generally located at and run by the institutions where the research is being done, whether universities, research institutes, or companies.

Human germline genome editing is clearly, at this point, human subjects research. Whether or not the edited embryo is a human subject, the prospective parents are, as is the woman who is to carry the pregnancy. That means in almost all countries with significant human medical research (and regulations), any efforts to make a baby through human germline genome editing will have to be reviewed by an IRB or its equivalent, will have to be judged to have potential benefits (to the subjects or to science or medicine) that justify their risks, and must provide for good informed consent. Germline genome editing—like any other human subjects research—will be subject to those rules and hence illegal unless it complies with these requirements.

It may seem I have spent too many words on this subject, but this human subjects research approval can be a major limitation on research. As discussed in detail in chapter 9, no respectable IRB in the United States would have approved He Jiankui's proposed research. For one thing, the risks grossly outweighed the benefits; for another, his consent form had major flaws. And if no IRB—or, in other countries, entity equivalent to an IRB (a research ethics committee in the United Kingdom, a research ethics board in Canada, and an IRB or ethics committee in China)—will allow this kind of research, it cannot legally be done.

The Law of Medical Product Approval

Some regulatory systems have additional barriers before some kinds of human subjects research can be undertaken. In the

United States, for example, although the FDA requires all U.S. trials to comply with the Common Rule's IRB system (and foreign trials to comply with something equivalent), it also has another type of approval required before a new drug can be used in human subjects research. As it happens, this system currently blocks any efforts in the United States to make babies through human germline genome editing.

The federal Food, Drug, and Cosmetic Act of 1938 (the FDCA), still the source of most of FDA's authority, generally prohibits the transportation or distribution of an unapproved drug across state lines.[10] We think of this prohibition mainly in connection with marketing and clinical use of a new drug, and, in fact, those uses are not allowed (in general) without approval by FDA of a "New Drug Application" (NDA). But to get an NDA approved, a firm will have to show that the drug is safe and effective for a particular disease or condition in humans. And in order to show that, it will have to give that experimental, nonapproved drug to humans. This would, without more, violate the act's prohibition on use of unapproved drugs.

And so there is more. The FDCA provides for something called an "Investigational New Drug exemption" (IND), which allows firms (and others) to do research with unapproved drugs for the purpose of collecting information that may lead to their approval.[11] An IND is not automatic; it must be requested. An investigator needs to submit an application that gives FDA evidence that the drug seems to be likely to be reasonably safe in humans and some reasons to think it may be effective. That evidence includes pharmacology and toxicology studies in nonhuman animals, studies of in vitro human cells or cell lines, results of past use in humans in foreign countries, information about how and how consistently the drug will be manufactured, and detailed protocols about the clinical trials as well as about the qualifications of the investigators. Once an investigator submits an IND application to FDA, it must wait at least 30 days to start the clinical trial. FDA has the power to block the IND during that time, putting a hold on it either permanently or pending new information. If it does not block it, the investigator may go forward.

For biopharmaceutical firms or medical research establishments, like medical schools, INDs are usually not hard to obtain. Generally, the sponsors of the trial will have had extensive discussions with FDA before submitting the IND application, learning just what kinds of information FDA will want to see in order to allow them to test the drug in humans. But, easy or hard, INDs are required for any U.S. research use of a nonapproved drug.

Since about 2000, FDA has asserted that any human embryo that been substantially modified, genetically or otherwise, is a drug or biological product, the clinical use of which requires FDA approval.[12] It first took this position in connection with human cloning and then extended it to mitochondrial transfer technology. That claim of jurisdiction has not been tested in court, at least about human embryos. My own belief is that FDA is very likely (but not certain) to prevail in any such lawsuit. As a result, clinical use of human germline genome editing (just like human somatic cell genome editing) would require a sponsor to have received FDA approval of an NDA (if it is viewed as a drug) or a Biological License Application (BLA) (if a biological product). Each would require lengthy, expensive, and painstaking proof that the process was safe and effective.[13] And even research use would require enough advance proof of safety to justify an IND.

It is clear that today FDA should (and would) block an IND for human germline genome editing purely on scientific grounds unless and until a great deal more safety information becomes available, both about its use in nonhuman animals and also about its observed effects on ex vivo human embryos. That is true without even considering the possible influence of political opposition to such efforts, opposition that can affect decisions by FDA or by the secretary of HHS, to which FDA reports.[14] Moving a modified embryo into a human uterus would then be considered the distribution of an unapproved drug or biological product, which is a violation of the FDA's defining statutes and can be punished with civil and criminal penalties. Assuming its assertion of jurisdiction over such embryos is correct, FDA's regulatory judgment is thus highly likely to block any legal use of human germline genome editing in the United States for many years even without any special legislation or regulation aimed at that technique.

But even though FDA almost certainly would not, any time soon, allow even an IND for germline editing to go into effect, Congress took preemptive action of its own to bind FDA's hands. In December 2015, it added an amendment to the legislation appropriating funds to FDA. Such amendments, typically added by committees near the end of the appropriations process, are referred to as "riders" and are usually not subject to committee hearings, discussion on the House or Senate floors, or any open discussion. They "ride along" with a crucial appropriations measure, one too important to be held up in an effort to strike the rider.

In relevant part, this rider said

none of the funds made available by this Act may be used to notify a sponsor or otherwise acknowledge receipt of a submission for an exemption for investigational use of a drug or biological product under section 505(i) of the Federal Food, Drug, and Cosmetic Act

(21 U.S.C. 355(i)) or section 351(a)(3) of the Public Health Service Act (42 U.S.C. 262(a)(3)) in research in which a human embryo is intentionally created or modified to include a heritable genetic modification. Any such submission shall be deemed to have not been received by the Secretary, and the exemption may not go into effect.[15]

I think I may bear some responsibility for this precise language. In June 2015 I talked to a reporter from *Nature* who asked me about congressional plans to try to block INDs for genome edited embryos through forbidding FDA from spending any funds considering an application for such an IND. I am quoted as saying "This step seems dumb—or ill-advised,"[16] and I have no reason to think the quotation is wrong. If, I told the reporter, FDA could spend no money on the application, it could not block it and thus the sponsor would automatically be allowed to proceed 30 days after FDA received the application.[17]

When the amendment was finally passed in December 2015, shortly after the first summit, its language said both that FDA could spend no money acknowledging receipt of an IND application and, more powerfully, that no such application could be "deemed" received. Sponsors can proceed 30 days after receipt, but if their application is never received, the 30 days can never run. And, under this rider, the application would not be considered received even if all nine members of the Supreme Court testified that they had seen it handed to an FDA receptionist. (Most likely someone else pointed this out to Republican staff members in the House of Representatives, so it probably isn't my fault. Still . . .)

Although acts appropriating funds are generally only binding for one year of appropriations, the rider, in the same language, has been renewed every year and remains in effect today. A similar appropriations rider, banning federal funding for any research that destroyed or threatened harm to any human embryo that was not directly aimed at treating that particular embryo, the so-called Dickey-Wicker Amendment, has been renewed every year since 1995.[18] This rider may not go away soon but instead could continue indefinitely to serve as a de facto moratorium on the use of germline genome editing in humans in the United States.[19]

In 2019 members of the newly Democratic majority of the relevant committee of the House of Representatives raised the possibility of eliminating the rider in favor of having hearings and other consideration of a more narrowly written version.[20] The push behind this seems to have been to consider eliminating the rider's effective ban on mitochondrial transfer research, an interesting technology that plausibly does make a "heritable genetic modification" by replacing a prospective mother's (unhealthy) mitochondria, which contain their own DNA, with healthy mitochondria from another woman.[21] This technique (unfortunately frequently known by the newspaper headline term, "three parent babies") is less controversial than human germline genome editing. It affects less than 0.001 percent of the baby's DNA, it does not use CRISPR or other "genome editing," and it is being explored legally in the United Kingdom. But after public Republican opposition, the committee quickly retreated and the rider was once again adopted.

So, to summarize for the United States: genome editing for human reproduction is only legal with FDA approval, either for research or for clinical use, but FDA is forbidden to consider or allow such use—and no application to FDA for an IND will even be considered "received," presumably no matter how many witnesses can swear that it was handed to an FDA official. And that is without any substantive ban on human germline genome editing—just the regulatory system's normal IND process, bolstered by a one-year-at-a-time congressional prohibition on letting that system consider such IND applications.

The Specific Law of Germline Genome Editing

Other countries may have choke points in their drug regulatory systems similar to the IND, although they may be less likely to treat edited human embryos as drugs. But, even long before the He Jiankui affair, many other countries had already passed substantive laws focusing on human germline genome editing—and banning it.

The United Kingdom did it 30 years ago. Its Human Fertilisation and Embryology Act of 1990 banned any uses of genome editing techniques in human embryos, eggs, or sperm intended for use in reproduction. It did so by limiting reproductive uses to "permitted embryos." A permitted embryo was one where "no nuclear or mitochondrial DNA of any cell of the embryo has been altered." Similar conditions apply to permitted eggs and sperm.[22] Use of unpermitted embryos in treatment is a crime. In February 2015, after a long period of study, Parliament approved an amendment to the act to allow the HFEA to license mitochondrial transfer.[23]

On the other hand, in vitro research use that does not involve the transfer of an embryo to a uterus for possible implantation, development, and birth is legal, if licensed by the HFEA.[24] In early 2016, the HFEA granted Dr. Kathy Niakan of the Francis Crick Institute a license with a provision that allows such research with genome edited embryos.[25]

Although the definitions and details vary, many other countries join the United Kingdom in having bans on germline modification in human reproduction. A 2016 publication counts at least 11 countries with express bans on the procedure, including Canada, Brazil, the United Kingdom, France, the Netherlands, Belgium, Germany, Israel, South Korea, Japan, and Australia.[26]

Several of those countries, and many more, are also required to ban human germline genome editing through something known as the Oviedo Convention. More precisely, it is the "Convention for the Protection of Human Rights and Dignity of the Human Being with regard to the Application of Biology and Medicine.[27] The convention is a product of the Council of Europe. This international organization is *not* part of the European Union. Founded in 1949, it predates even the earliest glimmers of today's European Union by several years. The two do share the same flag and anthem, and the Council of Europe is located in Strasbourg, home to the European Union's European parliament. It has 47 members, including almost every state in (and some just near) Europe,[28] while the European Union has only 27 (after Brexit). It has no power to make laws, but it can propose international agreements for European countries and can then enforce those agreements on members that have ratified them. Its best known body is the European Court of Human Rights, which enforces the European Convention on Human Rights.

In June 1990 the then–Secretary General of the Council of Europe proposed the creation of a framework convention on bioethics.[29] That started a more than six-year process of discussion, drafting, consultation, and amendment. The Committee of Ministers approved it in November 1996, and it was opened for signatures on April 4, 1997, in Oviedo, Spain. After Spain became the fifth country to ratify the convention, it came into force in December 1999. It has now been signed by 35 of 49 member states; 29 of them have ratified it; only these have implemented it into their national law. The United Kingdom did not sign the convention, viewing it as too restrictive; Germany did not sign it, viewing it as too lenient.[30] Among other large countries, Austria, Belgium, and Russia have not signed it. Italy, the Netherlands, Poland, Sweden, and Ukraine signed the convention but did not ratify it (and so are not bound by it).[31]

The convention covers many subjects, including, relevant to an earlier part of this chapter, ethical limits on human subjects research. The most important part here is Article 13 of the convention, which says,

> An intervention seeking to modify the human genome may only be undertaken for preventive, diagnostic or therapeutic purposes and only if its aim is not to introduce any modification in the genome of any descendants.

The Oviedo Convention does not provide any direct enforcement mechanism by the Council of Europe. For the most part, the Oviedo Convention can only be enforced in the courts of the countries that have ratified it. Under its Article 23, the ratifying states are required to "provide appropriate judicial protection to prevent or to put a stop to an unlawful infringement of the rights and principles" in the convention. Under Article 29, the European Court of Human Rights can be asked for an advisory opinion on the legal interpretation of the convention, but only by states that are party to the convention and the Council of Europe's Committee on Bioethics. To be heard at the Court, individuals would have to argue that the specific violations of the Oviedo Convention that they were alleging were also violations of the European Convention of Human Rights.[32]

What was the legal status of human germline genome editing in China before the He Jiankui affair? That answer is obscure. No clear prohibition, akin to those in the United Kingdom or under the Oviedo Convention, existed. Government guidance to Chinese in vitro fertilization (IVF) clinics, as revised in 2003, does appear to forbid his actions. A 2018 article, primarily about various national regulations of mitochondrial transfer technologies, has a table showing different countries' rules. For China, the table says that, in the 2003 Human Assisted Reproductive Technology Specifications, "Article 9 of chapter 3 prohibits genetically manipulating human gametes, zygotes and embryos for the purpose of reproduction."[33] A tweet from Antonio Regalado, citing to this same 2018 article, attaches a photograph of what he identifies as "the annex of a 2003 ministerial guidance to IVF clinics."[34] The photograph shows text in English, presumably a translation, which reads "(Third) the implementation of technical staff's code of conduct (Nine) prohibits reproductive purposes for human gametes, zygotes and embryos for gene manipulation."

I have not been able to find any further discussion of this provision. To me, at least, it seems very unclear, at least before the He Jiankui affair and the Chinese government's reaction to it, what legal force that guidance had and how it was to be enforced. (Since the He Jiankui affair, as will be seen below, the legality of this procedure in China seems to have been answered.)

And so, having defined our topic, described CRISPR, and discussed both the ethical debates and the legal status of human germline genome editing, it is now time to travel to Hong Kong in late November 2018 and to the revelation of He Jiankui's experiment.

II The Revelation and Its Aftermath

6 The He Experiment Revealed

The way public knowledge of He's experiment unfolded seems like something out of a novel—but a real-life novel that we can trace through Twitter, YouTube, and online posts. It is not clear to me how important the pathway of the revelation is, but it is fascinating and, I think, adds depth to our understanding of what He did, and why. This chapter looks at it as a play in three acts: before the Summit, outside Hong Kong; before the Summit, in Hong Kong; and at the Hong Kong Summit. Keep in mind, while reading it, that the twin pregnancy began in March and the babies were born sometime in October.

Just before the Summit—Outside Hong Kong

Antonio Regalado was the first to suggest that He was conducting unprecedented experiments. Regalado is an aggressive and enterprising science reporter for the *MIT Technology Review*, though I have always suspected that he would really like to be Woodward and Bernstein (both)—he obviously enjoys investigative reporting. On Sunday, November 25, at about 7:15 p.m. EST (8:15 the next morning, November 26, in Hong Kong), Regalado posted an article on the website of the *MIT Technology Review* entitled "Exclusive: Chinese Scientists Are Creating CRISPR Babies."[1] The article reported,

> According to Chinese medical documents posted online this month . . . a team at the Southern University of Science and Technology, in Shenzhen, has been recruiting couples in an effort to create the first gene-edited babies. They planned to eliminate a gene called CCR5 in hopes of rendering the offspring resistant to HIV, smallpox, and cholera.[2]

Regalado got his scoop by examining the WHO's Chinese Clinical Trial Registry (ChiCTR), the equivalent for China of clinicaltrials.gov, the NIH website in the United States that includes a list of all experimental trials in humans (whose sponsors submit them for listing). Regalado found a trial listed that was seeking volunteer couples to help create the first gene-edited babies, babies who would be resistant to HIV infection.[3] (According to the website, the trial had first been listed a few weeks earlier, on November 8, 2018.) Regalado had been following genome editing, and human genomics more broadly, in China for several years.

Regalado wrote,

> The clinical trial documents describe a study in which CRISPR is employed to modify human embryos before they are transferred into women's uteruses.
>
> The scientist behind the effort, He Jiankui, did not reply to a list of questions about whether the undertaking had produced a live birth. Reached by telephone, he declined to comment.
>
> However, data submitted as part of the trial listing shows that genetic tests have been carried out on fetuses as late as 24 weeks, or six months. It's not known if those pregnancies were terminated, carried to term, or are ongoing.[4]

Regalado's question was answered in about two and a half hours. At 9:48 EST that evening, *STAT* published a tweet linking to an AP story by Marilynn Marchione, entitled "Chinese Researcher Claims First Gene-Edited Babies."[5]

Marchione's story totaled over 1,700 words—so she clearly did not write it in the 153 minutes that had elapsed since Regalado's piece appeared. The article says that He revealed his work "Monday in Hong Kong to one of the organizers of an international conference on gene editing that is set to begin Tuesday, and earlier in exclusive interviews with The Associated Press."

The article does not state how much earlier, but the AP must have been working on the story for at least seven weeks. The first photograph in the piece is of He in a laboratory in Shenzhen, and the caption says it was taken on October 10. This is one of the six photographs and a nearly three-minute video in the article, all of He or of three of his Chinese coworkers on his project. The article lists three contributors in China to the article's research, and said it was part of an "Associated Press series produced in partnership with the Howard Hughes Medical Institute's Department of Science Education." Apparently, Regalado's scoop effectively forced AP to release the article. Ryan Ferrell, He's public relations employee, says he gave Marchione permission to release it; AP says it made its own decision. (Surprisingly, and perhaps disingenuously, the AP story never mentions Regalado's piece.)

What did the AP story say? Marchione led with this: "A Chinese researcher claims that he helped make the world's first genetically edited babies—twin girls born earlier this month whose DNA he said he altered. . . ."[6] (This story also puts the twins' birth in November instead of the apparently accurate October.) The AP story then included much of what we discussed in chapter 3. It does differ slightly from other accounts in its numbers, saying, "In all, 16 of 22 embryos were edited, and 11 embryos were used in six implant attempts before the twin pregnancy was achieved, He said."

According to the article, He claimed he had ethics approval from Shenzhen Harmonicare Women's and Children's Hospital, and neither his home institution nor one named hospital of the four hospitals involved (the other three unnamed) provided the embryos. The article quotes Dr. Liu Zhitong, identified as the head of Harmonicare's ethics panel, as saying that "we think this is ethical."[7]

The article also discusses Michael Deem, He's Ph.D. advisor from Rice University:

> The U.S. scientist who worked with him on this project after He returned to China was physics and bioengineering professor Michael Deem, who was his adviser at Rice in Houston. Deem also holds what he called "a small stake" in—and is on the scientific advisory boards of—He's two companies.

Marchione adds,

> The Rice scientist, Deem, said he was present in China when potential participants gave their consent and that he "absolutely" thinks they were able to understand the risks.

Deem told AP that he worked with He on vaccine research at Rice and considers the gene editing similar to a vaccine. "That might be a layman's way of describing it," he said.

The article quotes three other American scientists, Kiran Musunuru, Eric Topol, and George Church, to provide ethical assessments of the work. That part of the article begins, "Some scientists were astounded to hear of the claim and strongly condemned it." Musunuru, who appears in an AP video distributed with the story, and Topol were strongly opposed—Musunuru calls it "unconscionable," and Topol says, "far too premature." Church's position was not as clear:

However, one famed geneticist, Harvard University's George Church, defended attempting gene editing for HIV, which he called "a major and growing public health threat."

"I think it is justifiable," Church said of that goal.

The article later notes that Church, along with Musunuru, questioned He's decision to transfer one of the embryos into a uterus for possible implantation and birth because He already knew that that embryo's cells had both one edited and one unedited copy of *CCR5*.

The Marchione piece remains, along with the unpublished He manuscript, our best source for information about He's experiment and the only interview we know he gave about his work. We do not know the background to the story. AP clearly invested a lot of time and effort into the reporting over a period of at least seven weeks (from at least the time of the October 10 photograph) and had He's cooperation. In 2015 He had hired an American public relations firm, HDMZ. Ryan Ferrell was an HDMZ employee and worked with He. I presume Ferrell arranged the AP coverage by Marchione, with whom he had worked in the past. (Ferrell went to work full time for He in April 2018; he moved to Shenzhen in August. In 2019, he worked part-time for He's wife—or, at least, was being paid by He's wife.)[8]

Regalado and Marchione's pieces were just the start. Much more came out that (U.S.) Sunday night. At some point on November 25—at least one time stamp shows it at 9:48 EST, the same minute the AP article appeared online—the He lab posted five short videos on YouTube, four of them featuring He, who spoke (in English) about the gene-edited babies.[9] The fifth video was by narrated by Dr. Qin Jinzhou, the lab's embryologist, speaking in Chinese (with subtitles in Chinese and English), about the twins.[10] Like the AP article, these videos had clearly had been produced well in advance of Regalado's revelation. Interestingly, the AP story did not mention these videos.

According to an August 1, 2019, story in *Science*, Ferrell

> had helped He lay out a plan to go public a month or two after the summit, syncing a published paper with an exclusive given to an AP reporter Ferrell had worked with in the past. But the scheme had unraveled: "This was everything not to plan."[11]

Ferrell and He had been surprised by Regalado's story; Ferrell said he had not known that He had, 18 days earlier, posted the project description in ChiCTR. After Regalado's story came out, Ferrell says that He gave the AP permission to post their story, and Ferrell and He posted the five (unfinished) videos.

One more piece from outside Hong Kong needs to be added. Sometime on Monday, November 26, *The CRISPR Journal*, a relatively new journal published by Mary Ann Liebert, Inc., put out an article, with He as lead author, entitled "Draft Ethical Principles for Therapeutic Assisted Reproductive Technologies."[12] The *Journal* has told me that the He paper was received on November 5. They sent it out for expedited peer review in the hope of publishing it before the Hong Kong Summit but knew nothing of He's efforts to make babies until the story broke in the press. On November 18, the *Journal* accepted it in principle, subject to some revisions. They received the revisions the next day and accepted the manuscript, scheduling it for publication on Monday, November 26, EST (the time zone of their editorial office). On the day after publication (and after revelation of the experiment), the editor-in-chief, Rodolphe Barrangou, and the executive editor, Kevin Davies, wrote to He, requesting a revised Conflict of Interest form. The original form declared no conflicts.[13] The article was eventually retracted toward the end of February 2019 on the grounds that the authors had failed to properly disclose certain conflicts.[14]

In the article, He and his coauthors, who included public relations advisor Ferrell, said,[15]

[W]e have thought deeply about ethical foundations for regulation in discussions between researchers, patients and advocates, and ethicists both in China and abroad. These discussions lead us to propose that, at a minimum, five core principles should be addressed in a modernization of Chinese regulations—and indeed any country's guidelines or laws—permitting gene surgery for ART [Assisted Reproductive Technology]: (1) a clear social purpose, (2) impermissible uses, (3) rights after treatment, (4) the human spirit's transcendence of DNA, and (5) a special duty to reduce economic inequality.

A box in the paper explains these five "core principles," though using different terminology:

1. Mercy for families in need . . .[16]

   A broken gene, infertility, or a preventable disease should not extinguish life or undermine a loving couple's union. For a few families, early gene surgery may be the only viable way to heal a heritable disease and save a child from a lifetime of suffering.

2. Only for serious disease, never vanity . . .

   Gene surgery is a serious medical procedure that should never be used for aesthetics, enhancement, or sex selection purposes—or in any way that would compromise a child's welfare, joy, or free will. No one has a right to determine a child's genetics except to prevent disease. Gene surgery exposes a child to potential safety risks that can be permanent. Performing gene surgery is only permissible when the risks of the procedure are outweighed by a serious medical need.

3. Respect a child's autonomy . . .

   A life is more than our physical body and its DNA. After gene surgery, a child has equal rights to live life freely, to choose his or her occupation, to citizenship, and to privacy. No obligations exist to his or her parents or any organization, including paying for the procedure.

4. Genes do not define you . . .

Our DNA does not predetermine our purpose or what we could achieve. We flourish from our own hard work, nutrition, and support from society and our loved ones. Whatever our genes may be, we are equal in dignity and potential.

5. Everyone deserves freedom from genetic disease . . .

Wealth should not determine health. Organizations developing genetic cures have a deep moral obligation to serve families of every background.

Why this long discussion of how He's work came to be known? In part because of its intrinsic interest (to me, at least), but in part to wonder, just when *was* He planning to reveal the twins? Ferrell told *Science* that he had laid out a plan to go public one or two months after the Summit. According to the *Science* article,

> In a text message obtained by *Science*, He said he had discussed the timing with his wife and with Zhang, a lab member, and "Mayor Xie"—who two He intimates say is Xie Bingwen, a deputy mayor and director of science and technology in Shenzhen's Nanshan district. . . . "I decided to set the announcement date of birth at around Nov. 20," He wrote. That alarmed Ferrell and the lab, and at their request Quake intervened, as he explained to *The New York Times*, trying to persuade He to wait until he published a paper.[17]

A *New York Times* story based on an interview with Quake is even more interesting on this issue of timing. According to it, in late October (after learning in mid-October of the babies' birth) Quake first texted and then spoke on the phone with "an extremely prominent scientist in the field" about the births. Quake said he had written that person "Mums the word for a

few more weeks but I thought you would like to know."[18] The
story continues,

> About a week later [a week after "late October"], Dr. He's publicist,
> Ryan Ferrell, contacted Dr. Quake, worried that Dr. He presenting the
> project publicly so soon could cause "severe and permanent harms to
> his reputation and the field." And, "the twins are still in the hospital,
> so no positive imagery."

At some point, perhaps immediately or shortly afterward,
Quake, in Hong Kong on other matters, met with He and Ferrell. He says he "advised Dr. He to submit the research to a peer-
reviewed journal, and Dr. He did so." This, presumably, was early
November—which fits with the ChiCTR filing on November 8,
a necessary prerequisite to publication in most journals. (The
manuscript specifically notes the filing.) The *Times* story says,

> Then, because journal review takes time, Dr. Quake said he advised
> Dr. He not to go public in Hong Kong, but to speak privately with key
> experts there so they can "get socialized to what's coming and will be
> more likely to comment favorably on your work."
>
> But Dr. He was not persuaded. "I do not want to wait for 6 months
> or longer to announce the results, otherwise, people will say 'a Chi-
> nese scientist secretly hide the baby for 6 months.'"
>
> Dr. Quake pushed back: "It is prudent to let the peer review pro-
> cess follow its course."
>
> But Dr. He went forward with his Hong Kong talk.

As discussed more below, on Monday, November 26, He, in
talking to Doudna and others, claimed to be uncertain at that
point whether to discuss his work in his scheduled talk two days
later at the Summit. Robin Lovell-Badge, the organizing committee member who moderated the panel He presented on, said from
the podium at the Summit that he had reviewed He's submitted
slides before the Summit, which included only preclinical work

and did not refer to the babies.[19] Lovell-Badge later said that, as the moderator of He's session, he held a conference call with all the speakers in the session including He "shortly before the summit." Before the call, He had sent him a draft summary of his talk, which did not mention transfer of embryos for implantation, let alone babies.[20]

Whatever his plans *before* that Monday morning in Hong Kong, once the Regalado article, the AP story, the five YouTube videos, and his ethics article in *The CRISPR Journal* were out, and it was clear the world's press was following his claims closely, it is hard to believe that He did not expect to talk about the babies at the Summit. How could He possibly expect to give a talk at the Summit and *not* talk about the babies? He was certainly ready to talk about the babies at the proverbial moment's notice.

On balance, I'm inclined to agree with Kiran Musunuru when he said of He "I suspect he was planning to pull a Steve Jobs style 'one last thing' during his talk."[21] What more dramatic setting for a reveal could he possibly hope for than the Second International Summit on Human Genome Editing?

Just Before the Summit—in Hong Kong

The Summit was to begin on Tuesday morning, November 27; He was to talk on Wednesday, November 28. Lovell-Badge gives some background on the timing of and reason for He's speaking role:

> A month or so before the Second International Summit on Human Genome Editing took place, several of us on the organising committee heard rumours that He Jiankui . . . was using genome-editing techniques on human embryos for the purposes of trying to make children who would be resistant to infection by HIV. We knew that JK had presented relevant work, involving genome editing in mouse

and monkey embryos, at meetings over the last couple of years, and that he had also started using the methods on human embryonic stem cells and human embryos in culture. However, there was concern that he felt he was in a position to try things for real: to make genome-edited human babies. We also heard rumours that JK had obtained local ethics committee approval to go ahead. Therefore, when we were deciding on additional speakers for the summit, JK's name came up. Although we were aware that he had not published in this area, he had clearly been doing relevant research and we thought it might be useful for him to attend the summit where the science, safety, ethics and regulatory issues surrounding genome editing would be discussed. We sent him an invitation and he responded almost immediately to say that he would be very happy to present.[22]

As the organizing committee gathered in Hong Kong on Monday morning, they arrived to the news about He's experiment. For at least one of them, that news had come a little earlier.

Jennifer Doudna, a member of the Summit's organizing committee, said she first got word of He's experiment in an email from him, which she received on Thanksgiving Day, November 22, three days before the Regalado piece. The email had the subject line "Babies Born."[23] Doudna was quoted saying, "I was just horrified; I felt kind of physically sick."[24] And again, saying, "Honestly, I thought, 'This is fake, right? This is a joke,'" she recalls. "'Babies born.' Who puts that in a subject line of an email of that kind of import? It just seemed shocking, in a crazy, almost comedic, way."[25] Doudna says as a result she changed her travel plans and left a day earlier for Hong Kong.[26]

Anne-Marie Mazza, then Senior Director of the NASEM Committee on Science, Technology, and Law, was the NASEM staffer in charge of the Summit.[27] She had arrived in Hong Kong on the morning of Friday, November 23 (late Thursday night or early Friday morning for Doudna in California). She found an email from Doudna waiting for her, asking, urgently, if they could talk. Doudna told Mazza about He's email over the phone; Doudna then telephoned David Baltimore, the organizing committee's chair, who was in the U.S., and informed him. Over the next few days, Mazza, Doudna, Alta Charo (who was also in Hong Kong early), and, by phone, Baltimore, talked about how to handle the He news, including how to inform the other members of the organizing committee. Lovell-Badge, for example, says Mazza contacted him on late Sunday afternoon, November 25, asking for an urgent meeting.[28]

Doudna arrived in Hong Kong Monday morning, about the same time He Jiankui had made the 90-minute drive from his lab in Shenzhen:

> "The nanosecond I landed at the airport, I had just a ton of emails from JK, desperate: I have to talk to you right now, things have really gotten out of control," recalls Doudna. . . . She went to Le Méridien hotel, where He was also staying, and checked in without immediately replying. "He actually had somebody come and pound on my hotel door."[29]

Sometime that morning Doudna and Lovell-Badge met briefly with He in the hotel lobby. (Remember, the Regalado article broke at about 8:15 a.m. in Hong Kong and the AP story no later than 10:48 a.m.) According to the same article in *Science*,

> When Doudna finally sat down with He in the hotel lobby on the morning of 26 November, a few hours after the news of the babies broke, the Chinese biologist seemed surprised by the immediate, intense flood of attention and mounting criticism, she recalls. He even asked her whether he should discuss the gene-edited babies in his talk. "It was bizarre," she says. "He seemed so naïve."

STAT gives more of Doudna's response:

> Um, Doudna replied, you've dropped this shocking news on the world, right before our summit, and you're not planning to mention

it? He seemed surprised that she expected him to but agreed to have dinner with her and other members of the summit organizing committee that evening to talk it out.[30]

At some time that morning, after the initial lobby meeting with He, Lovell-Badge and some other early arriving organizing committee members met to discuss how to handle the He bombshell. Should He present his work, how could they handle the He affair without letting it overwhelm the rest of the meeting, and whether and how to deal with the press about it before the meeting?[31] They "collectively took the decision that, assuming JK was still willing to talk, he should be encouraged to present at the summit."[32]

Doudna, Lovell-Badge, Alta Charo, and Patrick Tam, an organizing committee member from Australia, met for dinner with He at the hotel on Monday night, November 26, to discuss his work. Shortly after the Summit *Science* reported,

On the eve of the International Summit on Human Genome Editing in Hong Kong, China, last week, He, a researcher at nearby Southern University of Science and Technology in Shenzhen, China, had dinner at the city's Le Méridien Cyberport with a few of the meeting's organizers. The news of He's claim had just broken, and shock waves were starting to reverberate. But the reports were still so fresh that the diners sat in the restaurant without being disturbed.

"He arrived almost defiant," says Jennifer Doudna, who did landmark CRISPR work at the University of California (UC), Berkeley. She and the other conference organizers politely asked He questions about the scientific details and rationale of his work, the permissions he had secured to conduct it, and how he recruited hopeful parents to participate and informed them about risks. He asked them whether his planned talk two days later should include data about the twin girls, who had a gene altered to make them resistant to HIV infection. "We were all like, 'Uh, yes,'" Doudna says.

After more than an hour of questioning, He had had enough. "He just seemed surprised that people were reacting negatively about this," Doudna says. "By the end of the dinner he was pretty upset and left quite abruptly."[33]

A later *Science* article says,

At a dinner with He later on 26 November, Doudna and other summit organizers lobbied for full disclosure. He agreed to describe his work in detail at his talk 2 days later, although he said he had received threatening text messages and had switched hotels for safety. Alta Charo, a bioethicist at the University of Wisconsin Law School in Madison, asked He whether he understood the importance of the principles spelled out in the two main documents that gave germline editing a yellow light of sorts: the 2017 NASEM [National Academies of Sciences, Engineering, and Medicine] report and a similar July 2018 report by the Nuffield Council on Bioethics in the United Kingdom.

"I absolutely feel like I complied with all the criteria," He said.

"That kind of rocked me back," Charo says.[34]

Ultimately, the organizing committee decided that He would give his scheduled talk on the second day of the Summit, Wednesday, November 28, as part of a panel called "Human Embryo Editing," moderated by Robin Lovell-Badge and including presentations from Kathy Niakan, Paula Amato, Maria Jasin, and Xingxu Huang, but as a split session. The first four would speak, answer questions, and leave the stage. Then He would appear and would be the last to speak. The organizing committee released a statement at around 1:00 p.m. EST on Monday, November 26. In Hong Kong, that was about 2:00 a.m., Tuesday, November 26 (the starting day for the Summit), presumably after the dinner and just a few hours before the meeting's start.[35]

The statement read,

On the eve of the Second International Summit on Human Genome Editing, we were informed of the birth of twins in China whose embryonic genomes had been edited. The researcher who led the work, He Jiankui, is scheduled to speak at the summit on Wednesday.

The criteria under which heritable genome-editing clinical trials could be deemed permissible have been the subject of much debate and discussion by many research groups. . . . Whether the clinical protocols that resulted in the births in China conformed with the guidance in these studies remains to be determined.

We hope that the dialogue at our summit further advances the world's understanding of the issues surrounding human genome editing. Our goal is to help ensure that human genome editing research be pursued responsibly, for the benefit of all society.[36]

One might ask about the fairness to the conference organizers of He's cat-and-mouse game over "would he/wouldn't he." Of course, one might also ask whether, once they learned of He's work, the conference organizers should have allowed such ethically questionable research to be presented at the Summit. Under the circumstances—where the world, and the organizing committee, knew very little about what had happened and He was already scheduled to talk—I think they made the right decision . . . but I could be wrong.

At the Summit

The Summit opened on Tuesday morning with the usual welcomes and charges from local dignitaries and organizers. Four panels—two on science, one on ethics, and one on law—followed that day until the meeting's 6:00 p.m. adjournment. None of the panels focused on the He experiment.

The "Human Embryo Editing" panel was the third session on the following day, Wednesday, November 28. This panel was livestreamed, in China and around the world, and it is said that more than a million people watched.[37] (I did, and, as it was recorded and is available online, you can, too.[38] The following discussion of the panel draws from my viewing of the livestream and of the recording.) Lovell-Badge moderated the panel. Drs. Niakan, Amato, Jasin, and Huang gave their presentations and took questions for the first hour and 15 minutes of the panel, which had been allocated a total of 90 minutes. At that point, these four speakers left the stage, and the moderator, Robin Lovell-Badge, implored the audience not to interrupt He—telling the audience that he, Lovell-Badge, had the right to cancel the session if there were too many interruptions, and reminding everyone of the Hong Kong University's long tradition of free speech.

Dr. He presented his results for about 20 minutes, starting with mouse work and then moving to the babies. Afterward, He was questioned onstage by Lovell-Badge and Dr. Matthew Porteus, another scientist-member of the organizing committee, for about 15 minutes. David Baltimore, the chair of the organizing committee, spoke for a few minutes, stressing the need for societal consensus, arguing that further research would be irresponsible, and decrying that the experiment was neither transparent nor medically necessary. Baltimore called this "a failure of self-regulation by the scientific community, because of a lack of transparency."[39] For the remaining 25 minutes, He spoke with the moderators and then took questions from the audience.

Lovell-Badge questioned the selection of CCR5 given how little we know about it—particularly given some research that indicated it could make the babies more susceptible to influenza, and other research suggesting that editing CCR5 could enhance cognitive abilities. Dr. He responded that the gene had been "studied for decades," and that he was against using editing for enhancement. Porteus asked how many women were part of He's "pipeline" for his experiment, which is how we know (or think we know) that eight women were selected and one dropped out. Dr. He explained that eight couples were selected, one dropped out, and for the remaining seven couples, 31 embryos were injected, of which 70 percent were edited. He also explained that the clinical trial had ended given the "current situation."

Porteus wanted to know how the trial and the consent process was designed. Dr. He referred to his Cold Spring presentation,[40] where he apparently got feedback and criticism from some attendees; he also spoke with "top ethicists in the United States," and had "a U.S. professor" and "a Chinese professor" review his consent, along with the four people on his team. According to He, he personally spent one hour and 10 minutes with each participant to explain the consent, after each participant had spent two hours with one of his team members. He was confident that the women were "very educated" and could understand the consents.

At this point, the moderators opened it up to questions from the general audience and from the media. David Liu from the Broad Institute questioned whether the experiment satisfied an "unmet medical need," since sperm-washing technology can prevent prenatal paternal transmission of HIV. Liu also asked about the role of scientists in making decisions for patients. Dr. He said he felt proud about what he had done, to help the children survive, since HIV is such a horrible affliction. When pressed by another audience member on the ethics of his experiment, He said he was showing compassion by using available technology to help people with genetic disease.

Porteus interjected to ask if there were more pregnancies, and He told him that there was one. In a particularly difficult moment toward the end of the session, when Dr. Jasin from

Sloan Kettering asked about the personal impact on Nana and Lulu and the family dynamics between them, given the disparate outcomes of the experiment, He explained that he wanted to give them "freedom of choice." But He did not know how to answer when Dr. Jasin pressed him to consider how the families and the children, themselves, would deal personally with the fact that their genes had been edited.

At the end of the panel, He said he did not anticipate such a strong reaction from the international community.

I recommend that you watch the video of He's presentation for yourselves. My own reaction shifted during it. At first, I was, in spite of myself, impressed with how straightforward and sincere He seemed. But, as his presentation went on, I began to notice more and more gaps in his analysis. And, as he struggled to respond to questions, I began to feel that he was well out of his depth—in much deeper waters than any reasonable person would have put himself.

After the audience questions, He left the stage and, shortly thereafter, left Hong Kong to return to mainland China. He cancelled his scheduled appearance at the Thursday panel. In the more than one year between then and when I wrote this chapter, he made no known substantive statements on the research.

As interesting as the revelations were the reactions to them—both the immediate reactions to the experiments and the reactions in the form of delayed revelations that oozed out about who knew what and when. This chapter looks first at the reactions around the world and then at those specifically in China. The next chapter looks at the secondary revelations.

Dr. He said that he did not anticipate a strong reaction to his experiment. He was wrong. How did the world react? Two sets of reactions are particularly interesting. One is from the scientific community around the world; the other is from voices in China.

World Reactions

Even before He's talk, the first AP article contained strongly worded comments on his work by three prominent scientists:

> Some scientists were astounded to hear of the claim and strongly condemned it. It's "unconscionable. . . . an experiment on human beings that is not morally or ethically defensible," said Dr. Kiran Musunuru, a University of Pennsylvania gene editing expert and editor of a genetics journal.

"This is far too premature," said Dr. Eric Topol, who heads the Scripps Research Translational Institute in California. "We're dealing with the operating instructions of a human being. It's a big deal."[1]

And, as noted above, George Church (who, as a result of his comments on drafts, had been one of the listed authors of the *Science* article that came out of the Napa meeting in spite of not having been at the meeting) gave at least qualified support for He's goal. The AP story set the pattern for comments after the presentation—everyone expressed opposition to He's work, except, to some extent, George Church.

At the end of the Summit, its organizing committee issued a 10-paragraph statement. It stated, "The organizing committee concludes that the scientific understanding and technical requirements for clinical practice remain too uncertain and the risks too great to permit clinical trials of germline editing at this time."[2] The statement continued,

At this summit we heard an unexpected and deeply disturbing claim that human embryos had been edited and implanted, resulting in a pregnancy and the birth of twins. We recommend an independent assessment to verify this claim and to ascertain whether the claimed DNA modifications have occurred. Even if the modifications are verified, the procedure was irresponsible and failed to conform with international norms. Its flaws include an inadequate medical indication, a poorly designed study protocol, a failure to meet ethical standards for protecting the welfare of research subjects, and a lack of transparency in the development, review, and conduct of the clinical procedures.

This language was restrained compared with the assessments of some critics. Ed Yong hit some of the high points in an article in *The Atlantic*:

The CRISPR pioneer Jennifer Doudna says she was "horrified," NIH Director Francis Collins said the experiment was "profoundly disturbing," and even Julian Savulescu, an ethicist who has described gene-editing research as "a moral necessity," described He's work as "monstrous."[3]

I was quoted (accurately) as saying, "This is criminally reckless and I unequivocally condemn the experiment,"[4] and as calling the work "Grossly premature and deeply unethical."[5]

George Church was almost alone on the other side. Church, an immensely creative scientist at Harvard Medical School who often takes controversial positions, defended He's work in comments to several news sources. In an interview with *Science*, he said,

> I'd just as well not hang myself out to dry with someone I barely know, but I feel an obligation to be balanced about it. I'm sitting in the middle and everyone else is so extreme that it makes me look like his buddy. He's just an acquaintance. But it seems like a bullying situation to me. The most serious thing I've heard is that he didn't do the paperwork right. He wouldn't be the first person who got the paperwork wrong. It's just that the stakes are higher. If it had gone south and someone had been damaged, maybe there would be some point. Like what happened with Jesse Gelsinger [who died in a 1999 gene therapy experiment]. But is this a Jesse Gelsinger or a Louise Brown [the first baby born through *in vitro* fertilization] event? That's probably what it boils down to.[6]

GEORGE CHURCH

I have a great deal of respect, and some affection, for George Church. At the same time, I have, from time to time, referred to him as "the mad scientist of our times." Church is an immensely talented and creative researcher. Born in 1954, he is now the Robert Winthrop Professor of Genetics at Harvard Medical School, Professor of Health Sciences and Technology at Harvard and MIT, and a founding member of the Wyss Institute for Biologically Inspired Engineering at Harvard. He received a Duke undergraduate degree in zoology and chemistry in two years but "was withdrawn" from Duke's graduate biochemistry program, apparently because of his neglect of coursework and other formalities in favor of his research. He enrolled in a Ph.D. program at Harvard and finished a Ph.D. in biochemistry and molecular biology under Nobel Prize–winning geneticist Wally Gilbert in 1984. He joined the Harvard faculty in 1986 and has been there ever since.

Church heads a very large laboratory at Harvard Medical School, which becomes involved in almost every exciting biosciences issue. Church also has many ties with biotech firms, about 30 of which he helped found. His Harvard website lists over 230 (I lost count) companies and noncompanies for whom he has held advisory roles, consulting roles, or that are investors in companies he founded.[7]

I think I first met George in February 2012 when his lab hosted a meeting I attended to talk about "de-extincting" the passenger pigeon (an effort that I, at least in general, support). He also has a role in the CRISPR story as one of the first researchers (perhaps the first) to show that CRISPR could be used to edit human cells.

George is a visionary thinker who genuinely wants to help humanity. He also is unafraid to say controversial or very futuristic things, in interviews and in his book, *Regenesis: How Synthetic Biology Will Reinvent Nature and Ourselves*, coauthored with Ed Regis.[8] His efforts toward reviving the woolly mammoth as well as his speculations in *Der Spiegel*, a leading German magazine, about how one could "revive" Neanderthals[9] are examples of his willingness to be controversial, as is his discussion of making vast changes in the human genome, eliminating the redundancy in the human genetic code, in order to reduce vulnerability to viruses.[10]

It may actually be relevant that he is six foot, five inches tall and imposing for his height, with white hair, a white mustache, and a very full white beard. He also has blue eyes that beg to be called "piercing" and a very prominent forehead, that, although not Neanderthal in its details, adds to the unusual impression he makes. So does his deep and calm voice. Church is a force of nature and in many ways the most unusual researcher I have met.

A few weeks later, on December 14, something close to an "official" voice of capital "S" Science weighed in, in *Science* magazine. Victor Dzau, the president of the U.S. National Academy of Medicine; Marcia J. McNutt, the president of the U.S. National Academy of Sciences; and Chunli Bai, president of the Chinese Academy of Sciences, published an editorial entitled "Wake-up Call from Hong Kong." In it, they said that

> the case highlights the urgent need to accelerate efforts to reach international agreement upon more specific criteria and standards that have to be met before human germline editing would be deemed permissible. Together, we call upon international academies to quickly convene international experts and stakeholders to produce an expedited report that will inform the development of these criteria and standards to which all genome editing in human embryos for reproductive purposes must conform, and to engage scientific bodies around the world in this effort.[11]

Chinese Reactions

It was not immediately clear how China, and the Chinese, would react. The first Chinese story on He's work trumpeted it as a great accomplishment of Chinese science, "a milestone accomplishment *China* has achieved in the area of gene-editing technologies."[12]

That mood quickly changed. The same day—November 26—the He Jiankui story broke, a group of 122 Chinese scientists and ethicists published a joint statement on WeChat, a Chinese messaging service, calling the work "madness" and demanding stronger rules against such research. It deserves to be quoted in its full, 316-word English translation:

> Regarding the recent news from domestic and foreign media on human embryo gene-editing and two babies born using CRISPR

technology, as rational human beings, with respect for scientific theories and concerns regarding the future scientific developments in China, our statement is as follows:

The bioethics approval for this so-called "study" was insufficient. We can only use the word "crazy" to describe the experiment conducted directly on human beings. We have much to debate inside the scientific community about the accuracy and off-target-effects brought by CRISPR. Any attempts to alter human embryos and make babies carry huge risks without strict examination beforehand.

It is scientifically possible, but scientists and medical experts have chosen not to use the technology on human beings because of uncertainties, risks, and most importantly, the ethical problems that follow. Such irreversible alterations on human genes will inevitably go into the human gene pool. We should have a thorough and in-depth discussion with scientists and people across the world about these potential effects. We cannot rule out the possibility that the babies, born using this technology, can be healthy for a period of time. But the potential risks and dangers brought along by the unjustified procedures, especially if such experiments carry on, are hard to measure.

At the same time, this is a strike at the reputation and development of China's science, especially in biomedical research. It's extremely unfair to most of the scientists and scholars who work hard to innovate and adhere to ethical guidelines.

We urge related regulatory departments and affiliated research institutes to establish laws and regulations on [gene editing], and conduct a full investigation. They should also reveal the findings to the public.

Pandora's Box has been opened. We need to close it before we lose our last chance. We as biomedical researchers strongly oppose and condemn any attempts on editing human embryo genes without scrutiny on ethics and safety![13]

Many other prominent Chinese scientists condemned the experiment the same day and shortly thereafter.[14]

The government immediately announced on November 26 that there would be an investigation and suggested several (fairly vague) regulations that He's work may have violated.[15] On November 29, the Vice Minister of Science and Technology called for the suspension of any work at He's lab.[16] The Vice Minister for Industry and Information Technology announced a "zero tolerance" policy and barred He from competing for an award for which he had been nominated.[17]

After He left the Summit, his location was unknown for some time.[18] While rumors flew, including one that he had been executed, he was seen at the end of December in an apartment building at his university, the Southern University of Science and Technology.[19] Many said he was under the equivalent of house arrest, although Stanford ethicist William Hurlbut says He assured him over the telephone that was not true and he was free to come and go.[20] In any event, He did not use whatever freedom he had to make any further public statements about his experiment or his situation. (And he has still not been seen in public since the Hong Kong Summit, unless one counts his December 30, 2019, trial, open only to a limited number of invited observers.)

Various commentaries proliferated through the eight weeks after the Hong Kong Summit, with different perspectives and different analyses, but almost never with new information—until Monday, January 21, at about 6:30 p.m. in China. At that time, Xinhua, the official Chinese news agency, posted a story on XinhuaNet, in English, entitled "Guangdong Releases Preliminary Investigation Result of Gene-Edited Babies."[21] Guangdong is the province that includes the city of Shenzhen, where He's university and the hospital that allegedly gave ethics permission for the experiment are located. The story was based on an interview by Xinhua with one of the investigation team's members.[22] I have already quoted the short article, under 350 words, in its entirety in chapter 1. I will just note here that it was highly critical of He. It said he "defied government bans and conducted the research in the pursuit of personal fame and gain"; "intentionally dodged supervision, raised funds and organized researchers on his own to carry out the human embryo gene-editing intended for reproduction, which is explicitly banned by relevant regulations"; used "a fake ethical review certificate"; and "will receive punishment according to laws and regulations."

As far as I can tell, the report itself has never been published, in English or in Chinese. For many months Chinese officials made no further statements about He Jiankui's likely fate—except that on the day that Xinhua article came out, the Southern University of Science and Technology announced that he had been fired.[23]

Then, more than 11 months after the story on the Guangdong investigation, with no advance public notice, China announced on December 30, 2019, that He and two of his colleagues had been tried, had pleaded guilty, had been found guilty, and had been sentenced by a court in Shenzhen.[24] I have been able to find three direct reports on the trial, all released by Xinhua, the government press agency, two in Chinese and one in English. Two (one in Chinese and one in English) are short reports; the third is a longer "perspective" piece, which contains some information not found in the shorter versions. They provide some new information but very little. There are also many Western stories about the trial, but, as far as I can tell, they provide no information beyond that from the original Xinhua sources.[25]

The longer, "perspective" piece said the trial was held on December 27 (the Friday before the Monday release of the announcements) and had been based on a prosecution started on July 31.[26] It said that the defendants had been found guilty as a result of, basically, unauthorized practice of medicine:

> The court found that none of the three persons including He Jiankui had obtained a doctor's qualification and were still engaged in a series of medical activities, in violation of national regulations such as the Law of the People's Republic of China on Practising Physicians and illegal medical practice.
>
> Article 336 of the Criminal Law of the People's Republic of China stipulates that a person who has not obtained a doctor's qualification for practicing medicine illegally has serious circumstances and shall be sentenced to fixed-term imprisonment of not more than three years, detention or control, and shall be imposed a single fine or a fine; If a person is in good health, he shall be sentenced to fixed-term imprisonment of not less than three years and not more than 10 years, and a fine . . .

Somewhat confusingly, though, the reports also cite other violations by the defendants. That same article says,

> According to the "Ethical Guiding Principles on Human Embryonic Stem Cell Research" jointly issued by the Ministry of Science and Technology and the Ministry of Health in 2003, human blastocysts that have been used for research cannot be implanted into the reproductive system of humans or any other animal. . . .

And one of the Xinhua releases says,

> He Jiankui and others forged ethical review materials and recruited men to carry out gene editing and assisted reproduction for multiple couples of HIV-infected persons. By means of impostor and concealing the truth, gene-edited embryos were passed through assisted reproduction technology by an unknown doctor.[27]

Whatever the exact nature of the charges against him (and I have been unable to find the equivalent of an indictment, information, or complaint setting them out, which would exist in a U.S. case), He was sentenced to three years in prison and fined three million yuan, about $430,000. Zhang Renli was sentenced to two years in prison and fined one million yuan; Qin Jinzhou received an 18-month sentence and a 500,000-yuan fine. (The Xinhua report says there was a two-year "reprieve" for the sentences, although it is not clear from the text whether it meant for both Zhang and Qin or just for Qin.[28]) Zhang and Qin were the people who actually injected the CRISPR reagents into human embryos; following He's orders Zhang also forged the ethics review documents and illegally procured the materials for the experiment. Qin's involvement reportedly came in May and June of 2018 in connection with two couples who went to Thailand for the procedures, which he performed, but that did not lead to pregnancies.[29]

Chinese television reported, "Due to the personal privacy of the persons involved, the court heard the case in private," although "[t]he defendants' family members, deputies to the National People's Congress, members of the CPPCC [the Chinese People's Political Consultative Conference], media reporters and representatives from all walks of life attended the verdict."[30]

The television report also said,

> According to the person in charge of the court, during the trial, the public prosecution agency produced evidence such as physical evidence, documentary evidence, witness testimony, appraisal opinions, inspection transcripts, audiovisual materials, and electronic data. The three defendants pleaded guilty in the court, and the defense lawyers appeared in court to defend the three defendants.

It is only in the reports of the sentencing that we found out that a third baby had been born, presumably from the pregnancy He reported at the Hong Kong Summit in November 2018, although there are absolutely no details given: "They implanted genetically-engineered embryos into the women's body [sic] and impregnated two of them who gave birth to three babies."[31] We have no idea what gene or genes were edited in the third baby, for what purpose, and with what success.

Interestingly, after the announcement of the verdict, He was publicly supported by two Westerners. Josiah Zayner, perhaps the most famous "do-it-yourself" gene editing enthusiast, published a piece in *STAT* defending He:

> When a human embryo being edited and implanted is no longer interesting enough for a news story, will we still view He Jiankui as a villain?
>
> I don't think we will. But even if we do, He Jiankui will be remembered and talked about more than any scientist of our day. Although that may seriously aggravate many scientists and bioethicists, I think he deserves that honor.[32]

I like Josiah, who I think is not nearly as "out there" as he sometimes sounds (and I hope that his own self-publicized DIY self-experimentation doesn't harm him). But I do think he is wrong on this. Even if human germline genome editing turns out to be a good idea, we should not forgive pioneers who were reckless with human lives.

William Hurlbut (of whom we will hear more in the next chapter), an academic working in bioethics at Stanford, is one of the people who knew about He's efforts to make babies before the public, and a He confidant (he said he talked on the telephone for several hours at a time with He once a week between the Hong Kong Summit and mid-January). Hurlbut expressed sympathy for He in the aftermath of the trial:

> "Sad story—everyone lost in this (JK, his family, his colleagues, and his country), but the one gain is that the world is awakened to the seriousness of our advancing genetic technologies. I feel sorry for JK's little family though—I warned him things could end this way, but

it was just too late," wrote bioethicist William Hurlbut at Stanford University, whom He consulted on the embryo-editing experiment.[33]

As far as I can tell, George Church made no public comment on the conviction and sentence.

So, in terms of Chinese reactions to the He experiments, the trial, conviction, and sentencing of He Jiankui and two of his associates may be the single strongest statement we have, however frustratingly short and vague. But even for what we have been told, it is also worth noting that the governments of China, and of Guangdong Province, had their own interests in how He's work, and its relationship to those governments, was portrayed. I do not think it is too cynical to suggest that the Chinese reports of the investigation (like any government's reports of potentially embarrassing situations) should be viewed with some skepticism—whether second- or thirdhand short summaries of unpublished investigative reports or discussions of nonpublic trials.

8 Who Knew What When? Revelations of Pre-Summit Knowledge

The reactions to He's revelations came suddenly; so did information about other scientists, mainly Americans, who, before the revelation, knew of He's baby-making activities but said, and did, nothing. This chapter starts by describing what we know about who He told what and when. It then looks particularly at the non-Chinese scientist who appears most seriously involved, Michael Deem of Rice University. It ends by describing some of the potential revelations (whether bombshells or squibs) that are likely outstanding.

Outside Academics Whom He Consulted

Since the revelation of the He experiment, several academics have come forward to say that they had had conversations with He in which he talked about his plans to edit human embryos for birth; some of them even knew that he had started pregnancies. Most say that they tried to dissuade him and did not know that he was planning to disregard that advice and proceed. None of these academics "informed" on him.

Quite quickly, we came to know of at least six such academics: scientists Mark DeWitt, Craig Mello, Stephen Quake, Matthew

Porteus, and Michael Deem, plus ethicist William Hurlbut. An August 1, 2019, article in *Science*, by Jon Cohen, with the eye-catching title "The Untold Story of the 'Circle of Trust' behind the World's First Gene-Edited Babies," added several more. At a Washington University meeting we both addressed in October 2019, Ben Hurlbut, a Science, Technology, and Society scholar at Arizona State University, put the number of academics who knew in advance at over 60. (I asked Ben to name names, citing, somewhat in jest, Senator Joseph McCarthy's mythical list of 205 known communists working at the State Department—a number that jumped around subsequently but eventually centered around 57[1]–but, fairly enough, he said I would have to wait until he had published.)

Let's start with six who were revealed quickly. A November 27 article in *STAT* discusses Mark DeWitt—a genome editing researcher at UC Berkeley. DeWitt had coffee with He in 2016, after He reached out to him.[2] They talked about technical problems in gene editing and stayed in touch, with DeWitt accepting an invitation from He to lecture at his university in January 2017. In September 2017, DeWitt says he was shocked when He told him in an email that he was starting a clinical trial to make edited babies. "I thought it was a terrible idea, with or without any kinds of approvals. I told him that. I said: 'You're not ready.'" According to DeWitt, they met again in person for dinner in January 2018, at which point DeWitt once again tried to convince He not to go forward. DeWitt continued to do so in further conversations.

Dr. He notified DeWitt in early November 2018 when the babies were born and sent DeWitt a description of the experiment, which He said he intended to submit to the *New England Journal of Medicine*. DeWitt told him the piece had substantial problems. At no point did DeWitt inform anyone else, telling *STAT*: "I wasn't sure what to do, frankly. He asked for confidentiality, and told me it was all above board on his end, so I let it be."

In 2006 Craig Mello and Andy Fire won the Nobel Prize for Physiology or Medicine for discovering RNA interference. Now a professor at the University of Massachusetts, Mello was at one point a member of the Scientific Advisory Board of one of He's companies, Direct Genomics. On January 29, 2019, the AP reported,

> Nobel Prize winner Craig Mello of the University of Massachusetts learned about the pregnancy last April from He in a message titled "Success!"
>
> "I'm glad for you, but I'd rather not be kept in the loop on this," Mello replied. "You are risking the health of the child you are editing . . . I just don't see why you are doing this. I wish your patient the best of luck for a healthy pregnancy."[3]

The story continued with more details:

> In April, He emailed Mello: "Good News . . . the pregnancy is confirmed!" He asked Mello to keep the news confidential.
>
> Mello, who won a Nobel in 2006 for genetics research, expressed concern about health risks. "I think you are taking a big risk and I do not want anyone to think that I approve of what you are doing," he wrote. "I'm sorry I cannot be more supportive of this effort, I know you mean well."

Mello resigned from the Scientific Advisory Board of the He-founded company, Direct Genomics, on December 6, 2018. There is, so far, no evidence that the company was involved in the He experiment (though sources of funding for He's work remain unclear). Nor is there any evidence that Mello tried to report He's work to anyone.

Stephen Quake, a Stanford bioengineering professor, supervised He's postdoc in 2011. Quake is a prodigiously creative researcher, who specializes in creating and using biomedical tools, especially for nucleic acid sequencing.[4] One of his interests is sequencing DNA reliably from single cells; this appears to have been what He worked on in his lab.[5] According to an AP story,

> Quake said he had met with He through the years whenever his former student was in town, and that He confided his interest a few years ago in editing embryos for live births to try to make them resistant to the AIDS virus.
>
> Quake said he gave He only general advice and encouraged him to talk with mainstream scientists, to choose situations where there's consensus that the risks are justified, to meet the highest ethics standards and to publish his results in a peer-reviewed journal.
>
> "My advice was very broad," Quake said.[6]

An article in the *MIT Technology Review* stated, "A much smaller number [of academics], including Stanford's Quake and Craig Mello . . . knew by the middle of last year that He had already established pregnancies."[7]

On April 14, 2019, the *New York Times* published a lengthy story on its front page about Quake, in which he was interviewed by *Times* reporter Pam Belluck after giving the *Times* access to all his relevant emails (although some had redactions).[8] The story is consistent with the AP story, though it provides a finer level of detail. Three of those details are particularly interesting. First, He emailed Quake in early April 2018 to inform him of the pregnancy. Second, He emailed Quake to tell him of the birth sometime in mid-October. And third, in late October Quake first texted and then had a telephone conversation with someone he described as "an extremely prominent scientist in the field," whom he also informed of the birth. Apart from that unnamed scientist, Quake does not appear to have informed anyone else. The article concludes quoting Quake: "To the extent that it wasn't obvious misconduct, what does a person in my position do? Encourage him to do it right, his research, right? I mean, that's what I believed I was doing."

STEVE QUAKE

I have known my Stanford colleague Steve Quake since at least 2008. He is one of the smartest and most creative scientists I have ever met. Born in 1969, he got his bachelor's degree in physics and a master's degree in mathematics at Stanford in 1991. He was a Marshall Scholar at Oxford, where he received a D.Phil. in physics in 1994. He did a postdoctoral fellowship with Stanford professor (and Nobel Prize winner) Steven Chu on single molecular biophysics. His first faculty position was at Caltech in 1995, when he was 26 years old. He eventually received a chair at Caltech but returned to Stanford in 2005 to help launch its new department of bioengineering. He is now Stanford's Lee Otterson Professor of Bioengineering and Applied Physics and an investigator of the Howard Hughes Medical Institute.

At 51, tall, and in the early stages of male pattern baldness, he has had a spectacularly successful career. In a rare trifecta, Quake has been elected to the U.S. National Academy of Sciences, National Academy of Engineering, and Institute of Medicine (now National Academy of Medicine). He has won a variety of international scientific prizes and has been involved in the founding of at least seven companies. Since 2016 he has been a copresident of the "BioHub," a collaboration between Stanford, UC San Francisco, and UC Berkeley that is supported by $600 million from Facebook's Mark Zuckerberg and his wife, Priscilla Chan.

Quake has made contributions in a variety of fields, but his core expertise is in dealing with very small biological quantities or entities. He has done important work in microfluidics and in single-molecule biophysics, including sequencing DNA from single cells. I was particularly impressed with how he used his technology for DNA sequencing to create noninvasive prenatal tests, based on fetal DNA found in the bloodstream of pregnant women.[9] I think of him as Stanford's non–"out there" version of George Church—unlike Church, I've found Quake very careful and deliberate in his statements about his science and its social implications.

I first got to know Steve well in 2009, after he had sequenced his entire genome, a rare event at that time. A group of Stanford faculty, led by Euan Ashley at the medical school but including me, took part in a medical analysis of Quake's genome, an effort in which Quake was both the "patient" and a coauthor.[10]

I am in awe of Steve's creativity and insights, and I like him. At the same time, I *don't* see Steve as someone who wears his heart on his sleeve and who immediately responds to questions with whatever is in the top of his mind. I think in his answers to questions, about He Jiankui and otherwise, he picks his words carefully.

Matthew Porteus[11] is a stem cell researcher at Stanford with a strong interest in using genome editing, including CRISPR, to treat people with sickle cell diseases and other genetic conditions.[12] He was also on the organizing committee for the Summit and, along with Robin Lovell-Badge, moderated the question-and-answer period following He's presentation. The publication *Xconomy* interviewed Porteus on December 4, shortly after the Summit:[13]

[Matthew Porteus]: About nine months ago, in February, JK [He's nickname] told me he was planning on doing this. His email said that he was in the Bay Area visiting with a graduate student of his, and they'd love to set up a time to talk.

[Xconomy]: So you met face to face?

MP: Yes. He started out on his non-human primate work, that he had modified embryos and attempted to implant them into animals but gotten no pregnancies. I was like, oh, thanks for the update. Then he said, now we'll start doing this in humans. That was shocking to me. I was totally blindsided.

I was more than chiding him. I was berating him. I told him he was putting the entire field at risk through his reckless actions. He was in what I thought was stunned silence. But he didn't try to defend himself. The graduate student was with him but didn't say anything to my recollection. I hadn't heard from him since.

Also interesting is Porteus's answer to a question about what he would do differently:

MP: Two things: One is call other people I knew he might have been speaking to, and as a group we might have come up with a decision. And perhaps I could have reached out for advice to someone more senior who has led study commissions and academies, who understands the sociology of science, and without revealing the confidence, run the situation by them and get their feedback.

David Baltimore's comment at the summit, that if you hear about something like this, you should tell someone—I'm not sure he's exactly right. I still believe we need to trust other people on the other side of the table. Trust that one side will behave responsibly, and that the other side will respect confidences. If we lose that, we'll have lost a lot. On other hand, in medicine we have patient-doctor confidentiality, but it's also codified that if a patient plans to hurt themselves or others, you have an obligation to disclose. Perhaps this fits that scenario where it's such an egregious overstepping of bounds that it's worth violating the unwritten culture. That's where it would have helped talking with somebody more senior and getting feedback. I needed to do more. In retrospect I wish I had done that.

X: But then what concrete steps could you have taken?

MP: I don't know. One could be to call reporters and describe to them the situation and let them investigate. Another is to overtly go public and shout it from the mountains. A third option would have been to directly go to people in China: I want you to be aware of a researcher in your country. JK claimed to me he had IRB approval [sign-off from a medical ethics committee]. And he still claims that. It's not clear if it's real. To me it was like OK, you have some IRB review, they signed off.

On January 17, 2019, Porteus, along with William Hurlbut, appeared at an event I moderated, organized by the Stanford Center for Law and the Biosciences, which I direct. The video of that event is available online.[14] Porteus's comments at it are consistent with what he told *Xconomy*.

William Hurlbut is an adjunct professor and senior research scholar in neurobiology at Stanford, where he received his M.D.[15] He has taught and worked in bioethics areas for many years, most notably as a member of President George W. Bush's President's Council on Bioethics, chaired (for most of its existence) by Leon Kass. He met He Jiankui at the conference at Berkeley on ethical issues in CRISPR that Hurlbut had organized with Jennifer Doudna in January 2017. They continued to speak, so much so that He listed Hurlbut in the acknowledgments to his July 2017 Cold Spring Harbor talk.

After the Hong Kong Summit, *STAT* quoted Hurlbut on several aspects of his relationship with He:[16]

"I knew where he was heading and tried to give him a sense of the practical and ethical implications," Hurlbut said. "But he kept returning to the good that could be done."

He was not fully transparent with Hurlbut. When the two spoke most recently, this fall, "JK did not tell me that he had established pregnancies," said Hurlbut, who believes He should have done so. "He didn't reveal to me what the state of his research was, though I suspected he had either pregnancies or born babies."

[. . .]

And why did He violate what many scientists consider basic research norms? "I can't get into his head, but he has a very earnest desire to move the science forward," Hurlbut said. "My overall feeling is that he's a well-meaning person who wants his effort to count for good."

Still, Hurlbut said, "I disagree with what he did."

Hurlbut's discussion at the Stanford Center for Law and the Biosciences event on January 17 was consistent with what he told *STAT* in late November. The most interesting addition was Hurlbut's comment that, after the Summit, he continued to have telephone conversations with He about once a week, for three or four hours at a time. Hurlbut apparently did not disclose his suspicions about He's work to anyone and has not, to my knowledge, publicly discussed his reasons.

Hurlbut seems to have become He's main advocate (or apologist). He continues to urge that He is being treated unfairly:

He was "thrown under the bus" by many people who once supported him. "Everyone ran for the exits, in both the U.S. and China. I think everybody would do better if they would just openly admit what they knew and what they did, and then collectively say, 'Well, people weren't clear what to do. We should all admit this is an unfamiliar terrain.'"[17]

As late as November 26, 2019, the AP said Hurlbut believed that "so much effort has been focused on demonizing He that it has distracted from how to move forward."[18]

The three Stanford academics, Quake, Porteus, and Hurlbut, were the subjects of an investigation by the university concerning their contacts with He.[19] On February 7, Regalado reported that a Stanford spokesperson had said, "Stanford is reviewing the circumstances around Dr. He's interactions with researchers at the university."[20]

Two days after the April 14 *New York Times* article on Quake, another *Times* article revealed that Stanford had exonerated all three: Quake, Porteus, and Hurlbut.[21] The newspaper had received a copy of a letter from Stanford to Quake to that effect. The *Times* article further said that "the investigation of Dr. Quake began after the president of Dr. He's Chinese university wrote letters alleging that Dr. Quake had helped with the project." (The August 2019 *Science* article said the SUSTech president "had tried to shift the blame" to Quake.[22]) That same day, Stanford issued its own short announcement of the results of the investigation.[23] That statement did not name the three academics, stated that the review had been conducted by a Stanford faculty member and an outside consultant (neither named[24]), and, rather like the Chinese authorities and the Guangdong investigation, did not make the actual report available.

In an earlier article on this topic, I listed Benjamin Hurlbut, William Hurlbut's son, as another person who had advance knowledge of He's plans.[25] He has told me that he did not know—and I believe him (he definitely had substantial conversations with He before the revelations, but I now see no evidence that he knew) from his talks with He, any conversations he had with his clearly in-the-know father, or otherwise—that He was actually proceeding with his plans.

THE HURLBUTS

William (commonly known as Bill) Hurlbut has been at Stanford for most of his life. He received an undergraduate degree from Stanford in 1968 and an M.D. in 1974 (although he never did a residency and was thus never a licensed physician). He also did postdoctoral work in both theology and bioethics. He was for many years an instructor in the Stanford Human Biology program before becoming an adjunct in the neurobiology department. Bill Hurlbut is a conservative on bioethical issues, with a strong Christian (or perhaps Judeo-Christian) orientation.[26] He is probably best known for his membership on George W. Bush's President's Council on Bioethics, long chaired by Leon Kass. I don't remember when I first met Bill Hurlbut, but it had to have been no later than the 1990s, as I began to work in bioethics-related issues. Whether for political, religious, or just personality reasons, my relationship with him has never been close but also has never been antagonistic.

His son, Ben Hurlbut, is in some ways very different. Ben is an associate professor in the School of Life Sciences at Arizona State University.[27] He is generally far to his father's left. Ben received his Ph.D. from Harvard in 2010 in the history of science with a focus on science and technology studies (STS), where he worked closely with STS legend Sheila Jasanoff. Like many STS scholars, he takes a critical view of modern science and its capitalist setting.[28] I find myself often partly agreeing and partly disagreeing with Ben; I do not share his deep suspicion of researchers, but I do often share his concerns about inadequate attention in the biosciences to social issues. I don't think we are close, but we have good, straightforward conversations.

It is not clear how often He consulted with the younger Hurlbut. It is clear Hurlbut did not approve of his work: "These two lives are now an experiment, a matter of scientific curiosity, which is an outrageous way to relate to human lives."[29] On the other hand, Hurlbut said about He, "[I]t's wrong to call him a rogue when he's acting in line with [a scientific culture] that puts a premium on provocative research, celebrity, national scientific competitiveness, and firsts."[30] (I had to point out to one reporter that this statement is not an endorsement by Hurlbut of He as much as it was an indictment of contemporary science.) Hurlbut has written (and is writing more) on the case. Along with Sheila Jasanoff and Krishanu Saha, he has written about planning to launch a "global observatory" on gene editing, an interdisciplinary and international committee focused on more than just the science—focused on "the limits and directions of research."[31]

The "Circle of Trust" article in *Science* reports that a much larger number of people knew about He's plans. The article asks whether "He hid his plans and deceived his colleagues and superiors, as many people have asserted?"[32] It continues,

> Because the Chinese government has revealed little and He is not talking, key questions about his actions are hard to answer. Many of his colleagues and confidants also ignored *Science*'s requests for interviews. But Ryan Ferrell, a public relations specialist He hired, has cataloged five dozen people who were not part of the study but knew or suspected what He was doing before it became public. Ferrell calls it He's circle of trust.

Ferrell, the major source for the article, may have cataloged 60 people who knew or suspected what He was doing, but I have never seen such a catalogue.[33] The *Science* article does not list anywhere close to 60, but it does add a few new names.

One is Bill Efcavitch. Efcavitch first met He when Efcavitch was head of research and development at a company called Helicos, a sequencing company cofounded by Steve Quake. Helicos went bankrupt in 2012, but He licensed its technology for his start-up, Direct Genomics. Efcavitch, now chief scientist for a small firm called Molecular Assemblies, was "an early confidant" of He and worried about the effect of He's experiments (which he expected to fail to lead to viable pregnancies) on He's reputation. "Why do you feel you need to take that kind of risk with your personal career?" According to the article, Efcavitch along with Quake and "several others linked to Direct Genomics" were told by He about the pregnancy. These included Mello, who was at the time a member of Direct Genomics's scientific advisory board.

Another new name is Steve Lombardi, described as a former CEO of Helicos who now runs a consulting business. In August 2017 He told him about his embryo editing plans and asked for his help in finding investors for "genetic medical tourism." Lombardi says he talked to several potential investors about He's ideas, but in January 2018, after He pulled out of a meeting, Lombardi seems to have pulled back.

The article also names John Zhang. Zhang is a reproductive biologist, born in China and educated there through his medical degree before receiving a Ph.D. in reproductive biology at Cambridge. He runs a fertility clinic in New York called New Hope Fertility Center. And he has made a reputation for himself as being on the cowboy edge of IVF technology.[34] Zhang tried to get around U.S. laws by taking eggs and sperm from a couple from the country of Jordan, performing the mitochondrial transfer procedure in the United States, and then transferring the embryo in Mexico. Fortunately, that seems to have led to a not-obviously-unhealthy baby, but he still received a reprimand from FDA. He has since been involved in setting up a company along with a Ukrainian clinic doing mitochondrial transfer. The company's goal is to market the service to American women who are willing to travel to Ukraine.[35]

The *Science* article reports that He and Zhang met in August 2018 in New York to talk about jointly creating a Chinese clinic. In October 2018 (about when Lulu and Nana were being born), Zhang met He in Hainan Province, a tropical island in southern

China, which is interested in becoming a center for reproductive tourism. (The article also recounts how Zhang conveniently forgot the details of his relationship with He until presented with them.) It is clear from the article that Zhang knew of He's plans. It is not clear whether he knew that He had already proceeded to pregnancies.

The article adds one last name, that of a Chinese stem cell scientist, Pei Duanqing.[36] Pei is a member of the Chinese Academy of Sciences, was one of the 22 authors of the National Academies' Valentine's Day report, and was a member of the organizing committee for the Hong Kong Summit. The article says,

> According to two sources who did not want to be named, He said he informed Pei a few months before the summit about the implanted, edited embryos—which Pei strongly criticized. Pei declined to discuss the matter, saying He's actions were the subject of an ongoing investigation.

The new names brought out in the "Circle of Trust" are thus Efcavitch, a biotech company scientist; Lombardi, a businessman; Zhang, an aggressive fertility clinic owner; and Pei. Pei, the only person on the list living in China, is the only internationally known and respected academic named, and the evidence for his involvement seems weak. We don't know who make up the other 56 people "cataloged" by Ferrell, but, unlike DeWitt, Mello, Quake, and Porteus, these four, with the possible exception of Pei, do not add much to an argument that Science knew.

We know at least some other respected science people must have known about the activity shortly before it was revealed. Dr. He did submit an article about the babies to *Nature* in mid-November. Their editors, and possibly peer reviewers for them, would have known for a week or two, though in the confidential context of reviewing submissions.

Another one of the scientists who knew in advance, but in a different way, is Kiran Musunuru.[37] Musunuru, a genetic cardiologist at the University of Pennsylvania's Medical School, learned of He's work early but for unusual reasons. He had met Marilynn Marchione of the AP in November 2017 at the annual "Scientific Sessions" of the American Heart Association. They hit it off immediately, but he did not meet her again for a year, until the next such American Heart Association meeting. As they talked at that meeting, Marchione quizzed him about mosaicism—and did so in the context of a conversation about CRISPR. Ultimately, she asked him if he would be willing to review for her, on a very confidential basis, an unpublished scientific manuscript related to gene editing. She said she would get him the manuscript the following Sunday, November 18.

His anticipation grew when, on Thursday, November 15, the Thursday before Thanksgiving, she asked him if he would be willing to do a video interview with AP about the (thus far unseen) manuscript. He agreed and waited, eagerly, to find out what this manuscript was. It didn't arrive as promised on Sunday, or on Monday morning, but it came to him as an email attachment on Monday afternoon, November 19.

Musunuru writes, "I will never forget the moment when I realized what the manuscript was about. My instant, involuntary response: 'Goddammit!'"[38] Musunuru says he was "shocked but not really that surprised. The pessimist in me knew that it was only a matter of time before somebody charged ahead and try [sic] to make gene-edited babies." Musunuru did think *CCR5* was a frivolous target, and he had no recollection of ever hearing about or knowing He Jiankui. As he pored over the paper, though, he realized that the babies faced serious health uncertainties and risks that He had not noticed or commented on, particularly as a result of mosaicism. He grew angrier and angrier—and feared that He would not talk about his work in Hong Kong and that he, Musunuru, would have to continue to keep his knowledge bottled up.

It would be unfair to accuse Musunuru, who received the manuscript from a reporter, with a requirement of confidentiality, six days before the revelation, of acting improperly in not speaking up. Ironically, though, Musunuru eventually realized that he might have had an earlier clue about He's plans. As a peer reviewer, Musunuru had recommended rejecting a paper on editing *PCSK9* in embryos for lack of originality. Only after the Summit did he realize that He was the last author on that paper. One of the other author's names looked familiar to him; when he checked his email, he discovered that person, from He's lab, had contacted him asking about his work with that *PCSK9*. But it would also be a harsh judge who would conclude that Musunuru should have remembered those minor contacts, let alone used them to conclude that He was (probably?) actually trying to make germline genome edited babies.[39]

Michael Deem

And, finally, we get to the one American scientist who holds a very special, though not yet entirely clear, place in this story: Michael W. Deem, the John W. Cox Professor of Biochemical and Genetic Engineering and Professor, Physics and Astronomy, at Rice University, as well as the Founding Director of Rice's Program in Systems, Synthetic, and Physical Biology.[40] Deem was He's Ph.D. advisor. The first AP story says that "the Rice scientist, Deem, said he was present in China when potential participants gave their consent and that he 'absolutely' thinks they were able to understand the risks."[41] A second AP story says,

One independent expert even questioned whether the claim could be a hoax. Deem, the Rice scientist who says he took part in the work, called that ridiculous.

"Of course the work occurred," Deem said. "I met the parents. I was there for the informed consent of the parents."[42]

That first, frank avowal of his participation in the He experiment—presumably made when he thought He (and he, Deem) would be heroes—will quite likely come back to haunt him. The complete story remains unknown, but much can be said of Deem's role in the He fiasco, thanks largely to the dogged and diligent reporting of STAT and of reporter Jane Qiu.[43]

After the news first broke, Rice released a statement that it "has begun a full investigation" of Deem's role and that it "had no knowledge of the work."[44, 45] The statement continued that, "regardless of where it was conducted, this work as described in press reports, violates scientific conduct guidelines." The AP quoted unnamed "officials at Rice" as saying the work "is inconsistent with ethical norms of the scientific community and Rice University."[46]

Deem's lawyers quickly denied that he had played a substantial role: "Michael does not do human research and he did not do human research on this project."[47] According to the Houston Chronicle, "[a]sked to square the seeming discrepancy [between what Deem told the AP and the lawyers' statement], Hennessy [one of his lawyers] said the statement is all Deem's lawyers want to say for now."

I think Deem's lawyers have a tough case: it looks to me as though Deem was heavily involved in the He experiment from early on. In a subsequent AP story, Deem defended He's actions, saying the research team did earlier experiments on animals, connecting himself, through his choice of pronoun, even more closely to He's research (though not, in this statement, the human subjects research): "We have multiple generations of animals that were genetically edited and produced viable offspring, and a lot of research on unintended effects on other genes."[48]

On December 10, *STAT* reported on a He paper, submitted apparently around September on preclinical work modifying the *PCSK9* gene in mouse, monkey, and human embryos, which listed Deem as one of the 14 authors.[49] More recently, *STAT* has reported that He submitted another preclinical paper in late November, this one to *Nature*, on editing the *CCR5* gene. For both of these papers, "Author Contributions" statements say that Deem designed the project and wrote and edited the manuscript. *STAT* was provided one of the papers, and a scientist with a copy of the other manuscript read aloud its author-contributions statement over the phone.[50]

As we have already seen, He appears to have submitted a manuscript on the twins to *Nature*. A *STAT* article says that He's *Nature* submission had 10 authors with He listed first and Deem listed as the last author, a position often used in scientific publications for the senior author.[51] The version I saw, however, lists He as the last author and Deem just before him, as ninth. So do the versions of the article of that title that Antonio Regalado discussed[52] and the copy of the article reviewed by Kiran.[53]

That *STAT* article went on to say,

A Chinese scientist who worked on the project said Deem was more than a bystander: Deem collaborated with He on the experiment and participated as a member of the research team during meetings with several volunteers in 2017 as they were recruited and went through the informed-consent process—a crucial component of a clinical trial. Deem helped to obtain the volunteers' consent, speaking with them through a translator, said the Chinese member of the team, who asked not to be identified because the person was not authorized to speak to a reporter.

This story also quotes Deem's lawyers:

> denying that he was "the lead, last, or corresponding author" on the
> paper submitted to *Nature*: "Michael Deem has done theoretical work
> on CRISPR in bacteria in the past, and he wrote a review article on
> the physics of CRISPR-Cas. But Dr. Deem has not designed, carried
> out, or executed studies or experiments related to CRISPR-Cas9 gene
> editing—something very different."

(The lawyers apparently did not say Deem was not number nine
out of ten.) The STAT article quotes lawyers as saying Deem "did
not authorize submission of manuscripts related to CCR5 or
PCSK9 with any journal," but goes on to say "they then acknowledged that Deem was listed as an author on all three gene-editing
papers and said he had had [*sic*] instructed the journals to remove
his name from all the manuscripts." In seeming contrast with
Deem's prior statements, they said, "Dr. Deem was not in China,
and he did not otherwise participate, when the parents of the
reported CCR5-edited children provided informed consent."

As a former litigator, I appreciate ways in which these lawyers might have chosen their words carefully. "Designed, carried
out, or executed studies or experiments" are not the only ways
to have participated in them. Deem might, for example, have
drafted the papers reporting on the experiment—a useful role
for (presumably) the only native English speaker among a group
of authors submitting to English-language journals. When data
analysis was required, he may also have participated in that. That
Deem did not expressly "authorize" submission of the manuscripts is not the same as saying they were submitted against
his wishes or even against his expectations. And Deem might
not have "participated" when the two parents of Nana and Lulu
gave consent but still, as the AP quoted him saying, may have
been present at consent sessions for *other* parents. In fact, some

of the information we have indicates that Deem was present for consent sessions in the summer of 2017. We do not know when Lulu and Nana's parents went through the consent process, but the embryo transfer did not take place until late March or early April 2018, making it likely their consent process came later.

This is speculation. But these issues will likely be important for Deem. If Deem is considered to have been an actual participant in this research, instead of just providing casual advice, there could be serious consequences for him. Although it appears none of the work took place at Rice, universities are responsible to the U.S. federal government for work done by their faculty, wherever that work was performed. Martin Cline, the first experimenter in gene therapy, was punished by his employer, UCLA, for conducting gene therapy trials over a summer in Israel and Italy, because he had not received permission from the UCLA IRB to do the work—in Israel, in Italy, or at UCLA.[54] If Deem is held to have conducted "human subjects research," without Rice IRB permission, even if the research were conducted in China, Rice could face sanctions from the federal government, and Rice has the power—and, to mollify the federal government, a strong incentive—to punish Deem.

It could be that Deem was not too worried about Rice's reaction. According to *STAT*, Deem gave a talk at the City University of Hong Kong in June 2018 as part of a job interview for the position of dean of its College of Engineering.[55] He was supposedly offered the job a few months later and had been scheduled to take office in January. Instead, the school has appointed an interim dean.[56]

Rice said it launched its investigation into Deem almost immediately after the news broke of the He experiment and his role in it. I wrote this more than a year later and last edited it over 18 months later. I am surprised that, as far as I can tell, Rice has said nothing further about Deem. I have to believe that this will change, and before you can read this chapter—but then I thought it would have changed long before now.

Remaining Potential Revelations and Sources

Much remains unknown or unverified about this whole saga. It is not clear how much of it we will ever know. Here are a few ways we *might* learn more.

The text of the Guangdong Province investigation into the He affair would be interesting, as, of course, would reports from other investigations that the Chinese authorities, at any level, conduct. I worry that the Chinese government (and, perhaps more importantly, its Communist Party) will have no interest in releasing, or perhaps even in conducting, further investigations. They have their villain, they have made regulatory changes in response to the He experiments . . . I suspect they will be happy if no further information emerges and the memory of the whole matter fades away. And, given the ability of the state and the party to control access to people, documents, and information from China—at least on issues they care about—I will be pleasantly shocked if any outside individuals or groups are able to learn much more about what actually happened. At the most, we may get some more information, from carefully vetted Chinese scientists and doctors, on the genomic status of the twins and their health.

We got a little new information from He's trial, notably confirmation of the birth of a third child. Even in the unlikely case that there were important revelations, unlikely as the defendants all pleaded guilty, we likely will never hear of them. The trial was closed; there is no reason to think the Chinese government will make more information about it public.

I would be very surprised if we hear anything from He Jiankui himself. That is not impossible, but he has not been heard from publicly since the Summit, and even the last report of Western-ers talking with him is from January 2019. He has now been sentenced to three years in prison; both during and after that time, I do not expect we will see another interview of him from the AP, let alone a memoir, whether original or "as told to." I would not trust anything he said, but even some more suspect information might be helpful. For similar reasons I doubt that any of his Chinese colleagues will be talking to the press unless they are doing so at the authorities' bidding and with words the authorities have approved.

Perhaps some brave Chinese ethicists, academics, or journal-ists will provide some more information from inside China about what happened and why. But, again, unless they are reporting what the government wants and has approved, they would have to be brave indeed. I am not worried that they would be tortured or executed, but the Chinese authorities have plenty of ways to make the lives of people in China who irritate them miserable.

We are likely, I suspect, to hear at least something more about Westerners who knew or should have known about the He experiment. Whether that total, when added to the 10 or so already known, will come anywhere close to Ryan Ferrell's cata-log of five dozen remains to be seen. But I do await with interest Ben Hurlbut's planned work on this.

It is possible that Rice's investigation into the role of Michael Deem will both be interesting and be made public or leak out (although Stanford's investigation into Quake, Porteus, and Hurlbut has not been released or leaked, a nondisclosure that rivals China's in its completeness, if not in its importance). It is also possible that Deem will contest Rice's findings, either with his own statements or in court. That might provide some more detail.

Another possible card is held by Ryan Ferrell, He's public relations specialist. If he were to cut ties to China, he might be able to sell his story, either to the press or as a book, for a useful amount. He says he was not involved with He until after the pregnancy was started, but that gives him about six months or so of involvement while interesting events were going on. Such an account seems to me the most likely discussion from someone involved, though I would not easily trust what he had to say.

We have learned a little more as a result of leakage of the manuscripts He submitted for possible publication before the Hong Kong Summit, some quickly but the most interesting more recently.

A December 10, 2018, *STAT* article reported that He had submitted a 55-page paper to "an international journal," describing genome editing of mouse, monkey, and human embryos intended to modify the *PCSK9* gene and thus confer resistance to heart disease.[57] The article reported that the (unnamed) journal sent the piece out for outside peer review on October 2, and that it listed He as the senior author. The paper had 13 other authors, including Michael Deem. The journal apparently rejected the paper around November 17. The *STAT* article reported that "two genome-editing experts who read it" had serious scientific and ethical problems with it. A later *STAT* article said that the journal was *Science Translational Medicine*, a "second label" of one of *Nature*'s biggest rivals among scientific journals, *Science*.[58] *STAT* also reported that, in late November 2018, He submitted another paper to *Nature* on his preclinical work, dealing not with *PSCK9* modifications but *CCR5* edits.[59] These two papers, if and when released or leaked, will not tell us much, if anything, about the genesis of Lulu or Nana but may help us assess better what He knew or should have known before he proceeded with human experiments.

But then, a year later, on December 3, 2019, Antonio Regalado published portions of He's submitted manuscript about the CRISPR'd babies, as well as comments from experts (including me) on it.[60] Regalado says that the manuscript was considered, and rejected, by both *Nature* and the *Journal of the American Medical Association*. The article is fascinating, both for what He said and, more importantly, for what he didn't say. I think it is fair to say that the submitted manuscript does not tell us much more about what He did, but, as the commenting experts point out, over and over again, more about what he did not consider seriously—and just how bad this research was.

So . . . we know little about what really happened in the He experiments and why, and we may never know much more. To quote the always relevant movie *The Princess Bride*:

> **Inigo Montoya:** Who are you?
>
> **Westley:** No one of consequence.
>
> **Inigo:** I must know.
>
> **Westley:** Get used to disappointment.
>
> **Inigo:** 'kay.[61]

Nevertheless, we do know that some people outside He's research group—whether 16 to 60 remains unclear—*did* know about his experiments and said nothing to the press or the authorities about them. That raises questions to be explored later in this book.

III Assessing and Responding to the He Experiment

9   Assessing the He Experiment

In this chapter I analyze (or, more accurately, criticize) the He experiment. I stand by the comments I made to the press just after the news broke: the experiment was "criminally reckless" as well as "grossly premature, and deeply unethical."[1] The little we've learned since then has only strengthened my views. This was an unethical, and terrifying, experiment and would have been even if it were not human germline genome editing. But it was, which made it even worse.

To me, the experiment has at least five major problems: a terrible risk/benefit ratio; very questionable consent; inappropriate approval processes; complete opacity; and, finally, the violation of what came as close as possible in the world of science to an international consensus against germline genome editing (at least, yet).

The Risk/Benefit Ratio

One of the two most basic rules for human subjects is that the likely benefits *must* justify the risks being taken. As discussed in chapter 4, the Nuremburg Code is arguably the foundational statement of human research ethics.[2] Its sixth principle is that "the degree of risk to be taken should never exceed that deter-mined by the humanitarian importance of the problem to be solved by the experiment." In addition, the second principle states, "The experiment should be such as to yield fruitful results for the good of society, unprocurable by other methods or means of study, and not random and unnecessary in nature." And the fifth principle states, "No experiment should be conducted where there is an a priori reason to believe that death or dis-abling injury will occur; except, perhaps, in those experiments where the experimental physicians also serve as subjects." Seven doctors were executed as a result of those trials for committing war crimes and crimes against humanity, as at least partially encapsulated in the Code.[3]

The Helsinki Declaration[4] does not have the force of law and, in any event, applies, by its own terms, only to physicians, but is widely accepted as a source of guidance. It has a section on "Risks, Burdens and Benefits" that includes three principles:

16. In medical practice and in medical research, most interventions involve risks and burdens.

    Medical research involving human subjects may only be con-ducted if the importance of the objective outweighs the risks and burdens to the research subjects.

17. All medical research involving human subjects must be preceded by careful assessment of predictable risks and burdens to the indi-viduals and groups involved in the research in comparison with foreseeable benefits to them and to other individuals or groups affected by the condition under investigation.

    Measures to minimise the risks must be implemented. The risks must be continuously monitored, assessed and documented by the researcher.

18. Physicians may not be involved in a research study involving human subjects unless they are confident that the risks have been adequately assessed and can be satisfactorily managed.

> When the risks are found to outweigh the potential benefits or when there is conclusive proof of definitive outcomes, physicians must assess whether to continue, modify or immediately stop the study.

The United States goes farther. The so-called Common Rule governing human subjects research emerged over the period from 1966 to 1991. This federal regulation *does* have the force of law (in the United States), and it embodies most American law about human subjects research. The Common Rule requires that (most) human subjects research be reviewed by an IRB that, among other things, can only approve research when

> (2) Risks to subjects are reasonable in relation to anticipated benefits, if any, to subjects, and the importance of the knowledge that may reasonably be expected to result. . . . The IRB should not consider possible long-range effects of applying knowledge gained in the research (*e.g.*, the possible effects of the research on public policy) as among those research risks that fall within the purview of its responsibility.[5]

It further requires that risks to subjects be minimized "[b]y using procedures that are consistent with sound research design and that do not unnecessarily expose subjects to risk."[6]

The U.S. Common Rule is also not binding international law—let alone binding law in China—but it is another good example of the international consensus for applying research ethics rules from the Nuremberg Code, the Helsinki Declaration, and elsewhere. And, as noted in chapter 5 its general approach is widely mirrored for clinical trials because of harmonization among drug approval agencies.

Under all of these standards, the He experiment grossly failed the risk/benefit requirements.

Anytime someone tries a new intervention in humans, it requires the most thorough preparation—and crossing fingers, throwing salt over the shoulder, knocking on wood, or whatever superstitions or prayers the researchers hope may avert harm. First use in humans of a new technology is *always* inherently dangerous. First use in human embryos of something as powerful as CRISPR is terrifying . . . and not to be done lightly. Something that changes genes at the stage of a very early embryo has risks during the crucial (and still largely mysterious) nine months of embryonic and fetal development, through any eventual person's lifetime, and potentially for the lifetimes of some or all of her or his descendants. In He's experiment, the risks fall into two groups: the general risks of using CRISPR in human embryos and the risks of the particular changes planned (or made).

We know some of the risks of CRISPR in embryos. The CRISPR mechanism may make changes in the wrong places, creating so-called off-target effects that can be dangerous.[7] Or it may make wrong changes in the right places, creating both risks of harm from the unexpected edits as well as losing the potential benefits from the right changes. He's own work is an example; it created unexpected changes in the *CCR5* gene and did not create the sought after 32-base-pair deletion. CRISPR may edit only some of the cells of the embryos, resulting in "mosaic" organisms, with some edited and some unedited cells, which will not reap all (or any) of the potential benefits of the edits.[8] And it may make some other, unexpected changes—there is evidence that, at least in some cases, CRISPR can cause large deletions or duplications in DNA, with unpredictable (but almost certainly not good) effects.[9]

We might have known more about some of these risks. He claimed, in his August 2017 Cold Spring Harbor Laboratories talk, to have done experiments on CRISPR with mouse and monkey embryos, as well as over 300 human embryos.[10] Although papers describing at least some of them appear to have been submitted for publication in fall 2019, shortly before the revelation of his babies, those results have never been published. And there has never been any explanation for the long gap between his talk and submission. In any event, the work with mouse, monkey, and human embryos could only have told us about the risks for the first few days after fertilization, before implantation in a uterus. This would provide some data on the genomic risks of off-target effects, large scale deletions, and so on, but nothing about whether the CRISPR method had any continuing effects through embryonic and fetal development. In any event, it would take hundreds of carefully monitored births, lives, and deaths of mice and monkeys to give some comfort about human safety for the process. He never even claimed to have done that.

We also know that there are some risks associated with the particular modifications attempted. He's goal was to disable the *CCR5* gene in the embryos. No functioning *CCR5* gene should mean no CCR5 receptor molecules on the surface of any of the cells of the embryo, fetus, and eventual person. CCR5 receptors are normally found on several kinds of white blood cells, not just T cells (famous as the main target of HIV) but also macrophages, dendritic cells, basophils, and microglia.[11] They are also reported to be present in some breast and prostate cancer cells.[12] All of these cells, except the cancer cells, play crucial roles in the immune system. It turns out that research not rarely finds people with totally disabled genes who can survive and even be healthy. The human body has a great deal of redundancy and resilience. It is somewhat surprising that we tolerate *CCR5* deletions since *CCR5* (or its close relative, *CCR2*) seems to occur in a large percentage of vertebrates (other than fish)—mammals, birds, reptiles, and amphibians)[13]—which at least implies it is doing something important. But there are many people in the world who have no functioning *CCR5* gene. The rate seems (as far as we know) to be highest in Northern Europeans, with one study finding the highest prevalence of 2.3 percent in the Faroe Islands.[14] Some of those are apparently healthy middle-aged and older people, so it is clear that a functioning *CCR5* gene is not essential for a reasonable quantity and quality of life.

But what other effects might the absence of a functional *CCR5* gene have? Does it increase prenatal losses? Infant mortality? Young adult deaths or disability? Other problems? There is some—a little—evidence that it increases the risks to a person of West Nile virus infection[15] and perhaps of influenza infection.[16] Influenza kills, according to recent estimates, between 291,000 and 646,000 people worldwide every year, compared with about 770,000 who die from HIV infection.[17] In China, influenza leads to hundreds of thousands of excess deaths each year[18] compared with around 20,000 to 30,000 deaths per year from HIV.[19]

The healthy adults known to have no functional *CCR5* gene are from Northern Europe. They have a different environment, different diet, different microbial exposures, and (to some extent) different variations in genes other than *CCR5* than people in China. Is that important? No one knows, including He Jiankui.[20]

Our knowledge of the effects of not having a functional *CCR5* gene is limited in still another way. The most common source of nonfunctional *CCR5* genes found naturally in humans is the *CCR5Δ32* version ("Δ" here stands for deletion).[21] He Jiankui did not make this 32-base-pair deletion in either of the twins, who had other, unpredicted, mutations in their *CCR5* genes. These seem very likely to make those genes nonfunctional, but do those particular mutations have the same safety profile as

CCR5Δ32? We have no idea. No one has seen, let alone tested, the mutations allegedly found in those girls.

Those are some known risks. There is also the possibility of unknown risks. But risks are not vetoes; they must be weighed against the benefits to the subjects and to science.

The main benefit to the possible future children is the possibility of being free from the fear of HIV infection. Note that this does not mean free from the fear of HIV infection as a result of conception from an HIV-positive father. That risk can be easily minimized, and probably completely avoided, by washing the seminal fluid away from the sperm. Sperm do not become infected with HIV, though the fluid may carry HIV virus particles. Separate the sperm from the fluid and then fertilize the egg— through IVF or through lower tech artificial insemination—and the risk of infection at conception disappears.[22] That was done with Lulu and Nana.

So, the relevant benefit to the twin girls is the diminution of the risk of becoming HIV infected at some time in the future, through unprotected sex or possibly intravenous drug use. That benefit is reduced to almost nothing in this case, however, for several reasons. First, the benefit (from avoiding the risk of HIV infection) is small, several decades away, and can be averted (already) in other ways. The general method used by He, creating nonfunctional *CCR5* gene variations, is known not to be complete protection against HIV infection (and death).[23] And no one knows whether the specific changes He actually made to the embryos' *CCR5* genes—which he knew about before transferring them to their mother's uterus for possible implantation, pregnancy, and birth—protect against HIV infection or progression.

For baby girls born in late 2018, the risk of HIV infection is small and far away. Only about one person in a thousand in

China is infected with HIV, a prevalence about 75 percent lower than that of the United States and substantially lower than that of many European countries.[24] And other preventive measures, such as safe sex or avoiding intravenous drug use, are readily available, with who knows how many more to come before the girls become sexually active, let alone before they become intravenous drug users (if ever). Their risk of HIV infection seems close to, if not quite, zero. The benefit of avoiding it is correspondingly low.

But it gets worse. Most of the commentators have assumed that people without CCR5 proteins on their T cells cannot become infected with HIV.[25] (In his manuscript, He at least qualified this, saying it conferred "natural resistance against prevalent HIV-1 strains."[26]) Most infections occur through HIV infecting CD4 positive T cells.[27] The virus attaches to the CD4 protein on the surface of these cells *and* to another cell surface protein. Most often the second protein is CCR5, but it can be another protein called CXCR4.[28] In addition, other immune system cells, dendritic cells, can be infected with HIV both through the CD4 and CCR5 route and another route that does not require CCR5.[29] The most common strains of HIV currently infecting humans, the R5 strains of HIV-1, usually need CCR5 receptors to infect both CD4 T cells and macrophages.[30] But another HIV-1 strain, X4 HIV-1, does not use CCR5 at all but instead uses CXCR4.[31] Some strains of HIV-1, dual-trophic HIV-1 strains, can use either CCR5 or CXCR4.[32] And HIV-2, less common and less deadly, has strains that make use of many receptor molecules, including CCR5 and CXCR4.[33]

Thus, the absence of CCR5 proteins does not prevent infection with HIV although it does greatly reduce (and possibly entirely prevent) infection of the CD4 T cells with the R5 HIV-1 strains. The X4 and dual-tropic strains can still infect both CD4 T cells and macrophages; even the R5 HIV-1 strains can infect dendritic cells, which can serve as a reservoir of HIV where the virus can exist, and possibly mutate, even when the CD4 T cells are not infected.

And we have good evidence for the fact that the absence of CCR5 proteins is not completely protective. In 2012, the HIV world was excited by "the Berlin patient," later identified as Timothy Brown.[34] Brown was HIV-infected but also developed acute myeloid leukemia, a blood cancer. His leukemia resisted drug treatment, and he needed a bone marrow transplant, substituting someone else's blood-forming cells for his own, which were producing tumor cells. The doctors in Berlin gave Brown two transplants, one in 2007 and one in 2008, from the same donor, a person who had two copies of *CCR5Δ32*, the mutation that He wanted to create in Lulu and Nana. Brown had a very rocky course of treatment, which included full body irradiation, but eventually the bone marrow transplants took, and Brown was ultimately able to stop taking antiretroviral drugs without again becoming HIV positive. He has generally been viewed as the only person to have been "cured" of HIV.

In early 2019, the Berlin patient's possible successor was announced, the (so far still anonymous) London patient.[35] This patient had Hodgkin's lymphoma, a different blood cancer, but, like Brown, had received a bone marrow transplant from a donor without functioning copies of the *CCR5* genes. The London patient received his bone marrow transplant in May 2016 and, at the time his case was published, had been off of antiretroviral drugs for over 18 months with no recurrence of HIV infection. The *Washington Post* reported that this case was hailed as proof that Brown was not a fluke.[36]

But this second "cure" did not come easy. The same *Washington Post* article noted, "Despite efforts to repeat the remarkable Berlin results, researchers had failed for a decade." The *New York Times* provided more details:

> Once it became clear that Mr. Brown was cured, scientists set out to duplicate his result with other cancer patients infected with H.I.V.
>
> In case after case, the virus came roaring back, often around nine months after the patients stopped taking antiretroviral drugs, or else the patients died of cancer. The failures left scientists wondering whether Mr. Brown's cure would remain a fluke.[37]

The *Times* continued,

> One important caveat to any such approach is that the patient would still be vulnerable to a form of H.I.V. called X4, which employs a different protein, CXCR4, to enter cells. "This is only going to work if someone has a virus that really only uses CCR5 for entry—and that's actually probably about 50 percent of the people who are living with H.I.V., if not less," said Dr. Timothy J. Henrich, an AIDS specialist at the University of California, San Francisco.
>
> Even if a person harbors only a small number of X4 viruses, they may multiply in the absence of competition from their viral cousins. There is at least one reported case of an individual who got a transplant from a delta 32 donor but later rebounded with the X4 virus. (As a precaution against X4, Mr. Brown is taking a daily pill to prevent H.I.V. infection.)

The *Nature* article reporting the London patient specifically discusses that "one reported case," the so-called "Essen patient":

> The only other case of an HIV-infected patient transplanted with CCR5Δ32/Δ32 cells who interrupted ART [antiretroviral therapy] was the 'Essen Patient'. In this case in which ART was interrupted one week before allo-HSCT [transplantation of blood-forming stem cells from another person], a rapid viral rebound of a pre-existing minority HIV-1 variant able to infect cells via the alternative CXCR4 co-receptor was observed three weeks later.[38]

The Essen patient later died from a recurrence of the lymphoma that had required the stem cell transplant.[39] As the wording of the quotation from the *Nature* article implies, other HIV-infected cancer patients have received *CCR5Δ32* transplants and survived, but they have continued on their antiretroviral therapy, so it is not known whether the drugs or the transplant has kept their HIV levels low.

The key takeaway is that *CCR5Δ32* does not guarantee immunity to HIV infection and possible death. It works for strains of HIV-1 that use only CCR5 as a cofactor along with CD4 to infect T cells. Some reporters noticed, or at least commented on, the importance of CXCR4, including early news stories from *The Atlantic*,[40] *Nature*,[41] and Bloomberg.[42] But those facts then largely disappeared, probably because "complete" protection against HIV infection made for a better story, a better dilemma, and a better thought experiment, even though that it did not match the reality of the situation for the two baby girls.

And, for one of the babies, it gets even worse. One of the slides He Jiankui showed in Hong Kong demonstrated that the twin he called Nana only had *CCR5* edited on one of her two chromosomes. (Musunuru points out that He had apparently mislabeled the slide and that this was actually Lulu.[43]) Many people, especially in Northern Europe, have one normal copy of *CCR5* and one copy of *CCR5Δ32*. There is some evidence that their disease progresses more slowly—but they still can become infected and die from HIV. And, as Musunuru points out, as today's antiretroviral treatments for HIV can effectively stop this progression, this is a very small benefit indeed.

That alone makes at least the transfer of her embryo into the mother's uterus for eventual birth deeply wrong. She had very close to no possible benefit to be weighed against unknown, and unknowable at the time, risks. As a matter of clinical use of IVF, the parents might have been entitled to elect to transfer such an embryo, but as a matter of human subjects research ethics, it was grossly wrong.

The other twin has such previously unknown variants for both of her copies of *CCR5*. These variants do seem likely to prevent production of functional CCR5 protein and hence seem likely to resist infection, but biology is complicated, and we do not know for sure. Of course, we could have gotten some good evidence about this question. A researcher could use CRISPR to edit human T cells in the laboratory (*not* in an embryo or a person) to have those abnormal *CCR5* variants. The researcher could then test in the laboratory whether and to what extent HIV-1, either the R5 strain or the X4 strains, infected them. That would have been a good step to take *before* making human babies with that DNA. The manuscript He submitted about the experiment says he was planning to look at these issues now. It is beyond "too bad" that He Jiankui did not stop to try it first. As it is, any possible benefits to Nana or Lulu from HIV resistance, let alone immunity, are deeply unclear.

Finally, one other possible benefit from editing *CCR5*, not mentioned by He, needs to be discussed. Antonio Regalado has suggested that by modifying *CCR5*, He might have inadvertently changed, and enhanced, the girls' brains.[44] Some research has shown that mice with less functional CCR5 had improved memories[45]; other new research showed that people with one copy of *CCR5Δ32* recovered more quickly from strokes.[46] It is worth noting both that stroke recovery is not the same as cognitive enhancement and that none of the stroke subjects had two nonfunctional copies of *CCR5*. The evidence of benefit in stroke recovery, if any, applies only to the baby with one functional and one (presumably) nonfunctional copy of *CCR5*.

Regalado says there is no evidence that He set out to modify the children's brains. Regalado contacted the key researchers on this topic, and He had never contacted them for information, as he had contacted researchers on HIV and *CCR5* and *PCSK9* and heart disease. And, Regalado reported, He himself said, in the question-and-answer period in Hong Kong, "I saw that paper, it needs more independent verification. . . . I am against using genome editing for enhancement." Regalado further quoted the reaction of one of the scientists studying brain effects of *CCR5*, Alcino Silva from UCLA:

> Could it be conceivable that at one point in the future we could increase the average IQ of the population? I would not be a scientist if I said no. The work in mice demonstrates the answer may be yes. . . . But mice are not people. We simply don't know what the consequences will be in mucking around. We are not ready for it yet.

(Silva also said that when he learned of the birth of twins, "My reaction was visceral repulsion and sadness."[47])

What about the benefits to science? The fact that apparently healthy children could be born after embryo editing is a novel finding of nonzero value. Given earlier successes with monkeys, as well as other mammals, it is not surprising—but neither was it certain. Even closely related species sometimes differ markedly in reproduction and early development.

But even that very small benefit cannot be used to counteract the risks. Remember, the Common Rule further requires that risks to subjects are minimized by "using procedures that are consistent with sound research design and that do not unnecessarily expose subjects to risk." The He experiment did not have a sound research design, at least in that there was no published or otherwise generally available evidence from nonhuman trials about either the safety of this kind of use of CRISPR, in general, or the safety of *CCR5* inactivation, in particular.

To me, the balance is easy to weigh. The risks to the babies who might be born from the embryos grossly outweighed the almost-zero benefits to them and the relatively small benefits to science. It seems to me that no reasonable reviewing body—or researcher, who primarily carries that particular ethical obligation—could find otherwise. (I don't think this necessarily accuses George Church of being unreasonable, because his statements, though clearly contrary to the general theme of the commentary on He's experiment, are not necessarily inconsistent with condemning these particular actions—though they clearly are not a full-throated denunciation.)

These considerations of risk/benefit balance become all the more important in the context of this particular research—research on embryos that are not yet persons and cannot consent. If a mentally competent adult suffering from a dreaded disease decides to try a very risky experimental treatment and things go wrong, at least that adult had made a voluntary, and presumably informed, choice to take those risks. In He's experiment, the prospective parents certainly bore some risks, but most of the risk directly fell on any embryos that became babies. Those embryos never got a chance to consent.

I do not want to belabor this. Neither I nor, I strongly suspect, you, dear reader, gave consent to being born—let alone to whom, where, and when you were born. We do, and must, allow children, fetuses, and embryos intended for implantation to be research subjects without their own personal informed consent in order to learn how best to treat them.

But the Common Rule has special requirements for such research. Its Subpart B applies to fetuses (defined as any product of conception after implantation) as well as pregnant women and neonates.[48] Subpart D applies specifically to research with children.[49] Neither applies to ex vivo embryos, but the reasons for the special requirements apply equally there. The provisions protecting embryos and children are similar but are spelled out most cleanly in Subpart D, about children, which provides substantially heightened protections. The details are long and tedious (nearly 700 words in an earlier draft); I will just say that the He experiment could not qualify under those regulations.

One other point is worth noting. In the United States the Common Rule expressly bans IRBs from "consider[ing] possible long-range effects of applying knowledge gained in the research . . . as among those research *risks* that fall within the purview of its responsibility"[50] [emphasis added]. That's the binding American law; it is not necessarily an appropriate ethical condition. I don't think anyone needs to reach this question in the case of the He experiment, because even on a restricted view of the risks, they hugely outweigh the benefits. But, certainly, many people do think the "possible long-range effects" of human germline genomic editing include very substantial risks.

So far I have discussed weighing the balance of risk and benefit before the start of the He experiment. But both the Helsinki Declaration and the U.S. Common Rule require that the risks to human subjects be minimized, which the Helsinki Declaration at least makes clear is a continuing duty (something I think is also true, but less explicit, in the Common Rule): "Measures to minimise the risks must be implemented. The risks must be continuously monitored, assessed and documented by the researcher."[51]

He Jiankui says that he examined the DNA of the embryos that became the twins before making a decision to transfer them into their mother's uterus. At that time, he knew that he had failed to make the change he wanted, from a normal *CCR5* gene to *CCR5Δ32*. So he knew then both that the risks would be higher, because the edited versions of the gene had not been seen in humans, and that the benefits would be more uncertain. Even in the unlikely event someone concluded the experiment's balance of risks and benefits had been appropriate in advance, once it became clear that the planned edits had not been made, that calculus would change. And the continuing duty to monitor and assess risks would have prohibited the transfer of those embryos for possible birth.

The Dubious Consent

Along with a favorable risk/benefit ratio, proper consent is at the top of the list of requirements for ethical human subjects research. He's experiment seems quite likely to have failed that as well, for several reasons.

First, the consent process itself was flawed. George Annas from Boston University described it as more of a "contract" than a "consent form."[52] Its very first sentence says, "The research team is launching an AIDS vaccine development project."[53] Although one can see a parallel between genomic editing to provide immunity to HIV infection and the injection of an actual vaccine, that's only an analogy. To call embryo editing, which had never been used before in an attempt to lead to human babies, a "vaccine," a ubiquitous and widely accepted procedure, is deeply misleading.

In translation, the whole form is 23 pages, although the last 13 are technical annexes of little value to potential patients. The form does discuss *CCR5* and gene editing on the first page, but in a highly technical manner. As the primary benefit to the parent, it says (optimistically), "This research project will likely help you produce HIV-resistant infants."

Article 2 of the consent form for the mothers details the risks to the woman of the medical procedures she will undergo. Article 3 largely disclaims responsibility for these risks, specifically disclaiming liability for any risks that the woman will become infected with HIV or any other infectious disease, or that the children will not be HIV resistant. The third paragraph of Article 3 contains the only discussion of the risk to any children of the gene editing process. The risks it discusses are limited, and its goal seems more to try to avoid liability than to inform the prospective parents:

> The primary risk of gene editing (DNA-targeted CRISPR-Cas9 endo-nuclease) is the off-target effect of generating extra DNA mutations at sites other than the intended target. This is due to that the technique can cause nonspecific cleavage, resulting in mutations in non-targeted genomic sites. PGD [preimplantation genetic diagnosis], whole genome-wide sequencing, amniocentesis and peripheral blood test of mothers in different stages of pregnancy after transplantation will minimize the possibility of substantial injury. Therefore, this project team is not responsible for the risk of off-target which is beyond the risk consequences of the existing medical science and technology.

Damningly, at no point in its 23 pages does the consent document note that this technique has never before been used to try to make human babies.

There are also other "unusual" provisions in the "consent form." The seventh paragraph in Article 2 says,

> Regarding the qualitative characterization of the project results, only the project team has the right of final interpretation and announcement to the public. Then you have NO right to explain and have NO right to announce the project or result information without permission. Violation of this will dealt as breach of contract and the volunteers need compensate for the damages (The specifics are defined in the liquidated damages cooperation agreement).

(The document makes a similar statement in Article 10.)

Prospective research volunteers are given a free chance to withdraw from the research only up to a point:

> After the embryo implantation in the first cycle of IVF until 28 days post-birth of the baby, if you decide to leave the study due to other reasons than the ones listed in Items 3 and 4 above,[54] you will need to pay back all the costs that the project team has paid for you. If the payment is not received within 10 calendar days from the issuance of the notification of violation by the project team, another 100,000 RMB [Chinese yuan] of fine will be charged.

Article 7, section 2, of the document estimates that the project team will have spent 280,000 RMB on each research couple, mainly for the medical procedures. This is about $40,000, a huge sum for most Chinese couples (or American couples for that matter) to pay even before the additional 100,000 RMB fine.

The document describes the research procedures as trade secrets, forbids the subjects from disclosing them, and says they may not "use the secret through reverse engineering." Needless to say, this is a *very* odd provision, in what is supposed to be a consent form, to ask a nonscientist, nonphysician, prospective mother to sign. It is, at the least, some evidence that these investigators were neophytes at human subjects research and didn't know what a consent form should look like.

Now, it is possible that the form did not accurately reflect the discussions held between the researchers and the prospective parents; maybe they got a fuller description. He said he spent an hour and 10 minutes[55] describing the procedures to at least some of the prospective parents. Michael Deem told the AP that he attended at least some of the consent sessions and that he was confident in the parents' ability to understand the risks.[56]

On the other hand, we have no other reports on what the consent sessions were like. At the Summit, He claimed that the prospective parents were all well-educated;[57] some dispute this.[58] It is clear that neither He nor Deem, nor, apparently, anyone else involved, had been well trained on the processes of informed consent.

And, perhaps most powerfully, another problem with consent has slowly emerged. Why would parents choose to do a "never before done in humans" gene edit to make their babies? He said at the Summit that parents with HIV, facing what is a strong stigma in China, were eager to avoid the possibility that, at some remote time in the future, their children could become infected.[59] It is possible that they mistakenly thought that only the gene editing could prevent the prenatal transmission of HIV to the children, even though sperm washing could accomplish the same end with no risks.

But there is another, chilling, possibility. It may be difficult, expensive, and perhaps illegal and impossible for many people in China who are HIV positive to use the kind of assisted reproductive services that could include sperm washing. It is not clear that these services are available for HIV-positive couples. The Xinhua report of the Guangdong Province investigation says that "HIV carriers are not allowed to have assisted reproduction."[60] So it may be illegal; if not illegal, it would certainly be very expensive. If such a couple wanted genetic children—a very strong drive in Chinese culture—they could conceive the old-fashioned way, but that would risk prenatal transmission from the father to their children. Or, if they had a lot of money, they could seek IVF, though this would possibly be illegal. *Or* they could accept the offer to join He's experiment and not only receive IVF but receive it for free and receive other reimbursements. Dr. Kathy

Niakan, one of the researchers who spoke right before He in Hong Kong, said, "Offering vulnerable patients free IVF treatment presents a clear conflict of interest."[61]

This looks like "undue inducement"—the notion that people should not be lured, bribed, or coerced into taking part in research. "Undue inducement," as a concept, has been criticized, with some justice.[62] But certainly "take part in my experiment or I'll kill your child" is unethically coercive. When the inducement is "take part in my experiment as it is your only hope of having children without having a high risk of infecting them with HIV," the idea that the inducement is "undue" seems quite strong. This may be viewed as particularly true in a culture, like Chinese traditional culture, where having children to carry on the family (and especially boys to carry on the family name) is very highly desired.

The Approval Process

As to the approval process, it is not clear whether He got any approvals from *anyone* for his experiment. His own university, the Southern University of Science and Technology, says it knew nothing about the experiment,[63] which He echoed at the Summit.[64] It appears that the assisted reproductive centers, seemingly at four different hospitals, also claim not to have known what he was doing or to have approved it.[65]

He said the Shenzhen Harmonicare Women's and Children's Hospital provided the ethics committee, the equivalent of an IRB, that had approved the project. The original AP article quoted Lin Zhitong, whom it described as "a Harmonicare administrator who heads the ethics panel," as saying, "We think this is ethical."[66]

On the other hand, after the revelations, that hospital released a statement indicating that "the Medical Ethics Committee never met to discuss such a project, and that the signatures on He's approval form 'are suspected to have been forged.'"[67] In the same statement, Harmonicare said it "will invite public-security organizations to participate in the investigations and pursue the legal responsibilities of the relevant individuals."[68] The January Xinhua article on the Guangdong provincial investigation stated that He recruited subjects "with a fake ethical review certificate."[69] It is, of course, possible that the hospital is trying to cover up its role and the investigation did not detect, or even allow, the deception—we have little clearly reliable evidence for this, as for so much of the He Jiankui story.

Even more interestingly, the Xinhua article reports that "[A]s HIV carriers are not allowed to have assisted reproduction, He asked others to replace the volunteers to take blood."[70] If, as the court found at his trial, He violated Chinese law by procuring assisted reproduction for HIV carriers through recruiting uninfected men who would then provide blood for testing while falsely claiming to be the research subjects, that kind of fraud on an approval process should be actionable anywhere. The rule He violated may be unjust, but that does not excuse his fraud.

Transparency (or Opacity)

The guidelines and statements about human germline genome editing have universally called for transparency, broad discussion, creation of forums, and so on. Dr. He did absolutely the opposite. He created absolutely no public discussion of his experiment before the babies' existence was announced and acted to prevent any public knowledge of it. We do know that he talked to several people about the work—including at least Matthew Porteus, William Hurlbut, Steve Quake, Craig Mello, Mark DeWitt, Bill Efcavitch, Steve Lombardi, John Zhang, his apparent collaborator, Michael Deem, and perhaps Pei Duanqing. But many of them have said that He told them he was thinking about that kind of work, not that he was planning to do it, let alone that he had actually done it. Several advised He not to proceed, at least at this time.[71]

The robust discussion envisioned by various committees and others was nowhere to be seen. Although He worked in secrecy, he clearly was planning about how to disclose his work in order to make world news. He hired an American public relations specialist, Ryan Ferrell, who arranged for the AP to interview He in depth. Five YouTube videos were ready to put on the Internet on November 25, when the Regalado article first breached his cone of silence. Dr. He was not trying to avoid discussion, but he was trying to avoid discussion *before* his experiment was ready to reveal, discussion that might have interfered with his research and then with his ability to control the subsequent discussion of his work.

International (Near) Consensus

Finally, He violated what is as close to an international scientific consensus as I can ever remember seeing. Statements from organizing committees, national academies, and nonprofit bioethics organizations are not law. But they can represent an important consensus. The U.S. Declaration of Independence frames itself as something required by "a decent respect to the opinions of mankind."[72] The "opinions of mankind" were well represented in various statements before He's announcement about human germline genome editing. One should not fly in their face without good reasons. But, for reasons that seem focused on his own future fame, He did.

The most relevant statements are the early *Science* and *Nature* opinion pieces; the summary statement from the first International Summit; and, most importantly, the February 2017 National Academies report and the July 2018 Nuffield Council report. (By the time of the Nuffield report the pregnancy that led to Lulu and Nana had already begun; not so for the other documents.) The *Nature* piece flatly called for a ban on human germline genome editing. The others did not call for a ban but either said any such efforts today were premature or set out criteria that would need to be met before CRISPRing babies should be tried—criteria that were far from reality. He's experiment disregarded them all.

Consider the two most thorough and formal statements—the reports of the National Academies and of the Nuffield Council. The first listed 10 requirements, which we have already seen in chapter 4:

1. absence of reasonable alternatives;
2. restriction to preventing a serious disease or condition;
3. restriction to editing genes that have been convincingly demonstrated to cause or to strongly predispose to the disease or condition;
4. restriction to converting such genes to versions that are prevalent in the population and are known to be associated with ordinary health with little or no evidence of adverse effects;
5. availability of credible preclinical and/or clinical data on risks and potential health benefits of the procedures;
6. ongoing, rigorous oversight during clinical trials of the effects of the procedure on the health and safety of the research participants;

7. comprehensive plans for long-term, multigenerational follow-up while still respecting personal autonomy;

8. maximum transparency consistent with patient privacy;

9. continued reassessment of both health and societal benefits and risks, with broad ongoing participation and input by the public; and

10. reliable oversight mechanisms to prevent extension to uses other than preventing a serious disease or condition.[73]

HIV infection must be counted as a "serious disease or condition," although not nearly as serious as it used to be. That is, He's experiment may have satisfied the second requirement. But it clearly violated requirements number 4, 5, 8, 9, and 10. I would argue it also violated 1 and 3, and He gave no signs of having adhered, or of planning to adhere, to numbers 6 and 7. The report argued that "broad participation and input by the public and ongoing reassessment of both health and societal benefits and risks are particularly critical conditions for approval of clinical trials." He not only failed to meet this condition, but his secrecy prevented any possibility of the participation and input called for.

He's experiment similarly failed to meet the Nuffield Council's requirements. Most notably, it went forward with no wisp of the regulatory process the council sought:

> We recommend that heritable genome editing interventions should only be licensed on a case-by-case basis subject to: assessment of the risks of adverse clinical outcomes for the future person by a national competent authority (in the UK, the HFEA); and strict regulation and oversight, including long-term monitoring of the effects on individuals and social impacts.[74]

Of course, He could claim that he disagrees with those conditions (although, in fact, he has argued that he complied with them). But in his "ethics article" in *The CRISPR Journal*, published on November 26, 2018, he set out his own, largely vague, core principles. However, the experiments whose results he announced on November 25 on YouTube and on November 28 at the Summit violated the rules in that very paper.

In one concrete principle, He and his coauthors stated that "we hold additional but less universal beliefs that further restrict the use of gene surgery, including . . . focus only on treating disease via prevalent, natural genetic variants." This sounds to me like a restriction of "gene surgery" to cases where the (rare) pathogenic variant, say the dangerous *182delAG* mutation in *BRCA1*, is changed to the widely prevalent, "normal" variant, thus avoiding the risks of a new genetic variation. The 32-base-pair deletion in *CCR5* that He attempted to create in the embryos, while "natural," is nowhere "prevalent" and is quite rare, if not unknown, in China. But even if it does qualify, that would not justify He in transferring to a uterus the two embryos he did—one that did not have 32-base-pair deletion but has other mutations and one that had it only in some of its cells.

In the second of his five "core principles," called "Only for Serious Disease, Never Vanity," He and coauthors state, "Performing gene surgery is only permissible when the risks of the procedure are outweighed by a serious medical need."[75] People often will believe what they want to believe. It is possible that He genuinely thought his experiment met that requirement. I do not see how any reasonable person could agree that the risks of doing a "first in human" embryonic gene editing were outweighed by lowering the resulting baby's eventual risk of becoming infected with HIV, but, as Paul Simon wrote, "All lies and jest; Still, a man hears what he wants to hear and disregards the rest."[76]

10 Responses[1]

The He experiment was a fiasco. We can only hope that the two babies already born do not suffer from it, either in their health or from their possible resulting notoriety. More broadly, the experience has important lessons for Science as well as for China. Science does not have a president, prime minister, or pope. But Science does have leaders, individual and institutional, and those leaders have some influence over public perceptions. Leaders reacted—but their reactions were insufficient. Although Science is beginning to deal better, in some respects, with the consequences of He's experiment, its leaders still badly need to do better. China, on the other hand, does have leaders, and they need to take some specific steps—and, unlike Science, China's leaders seem to be well on the way to doing so.

The Inadequate Responses of Science

Big Science is beginning to move toward responding to the He Jiankui experiment, but it needs to do three things: enforce deterrence, create disclosure, and express humility.

Deterrence

He Jiankui expected to be hailed as a hero, or at least as an important pioneering figure. He gambled his high-flying present for the hopes of an even higher-flying future. He seems to have bet wrong. Far from being a hero, he has been (almost) universally condemned and has been tried, convicted, and sentenced to three years in prison. But whatever happened in China, Science needs to ensure he is ostracized. No future ambitious scientist should see this kind of experiment as anything but a suicidal career move.

In 1980 when UCLA's Martin Cline violated ethical rules by pursuing the first (unapproved) gene therapy trials, he lost positions and grants; his career never recovered.[2] Hwang Woo-Suk acted much worse by fraudulently claiming to have cloned human embryos. Until his fraud was discovered in late 2005, Hwang was a hero in South Korea, where his face graced a postage stamp. Following the revelations, Hwang was fired from his faculty position at Seoul National University, South Korea's premier research institution; he lost all his grants; and, in 2009, he was convicted of fraud and embezzlement and given a two-year prison sentence (suspended and later reduced to 18 months). Hwang has subsequently begun to rebuild his reputation with animal cloning work, but he has never regained his previous position.[3] Similarly, He Jiankui's career needs to be ruined—not necessarily forever, but for a long, long time.

How should Science accomplish this? Colleagues should shun him, journals should refuse to accept papers where he is an author, funders should forsake him. He needs to be on publicly announced blacklists, at the very least by journals and funders. And leaders of Science need to take the lead in announcing this and in encouraging others to do the same.[4]

Of course, collective ostracism, particularly coming from official or semi-official leadership, could descend into the abyss of McCarthyism or even Stalin-era Lysenkoism.[5] Individual scientists may decide, based on their own consciences, whether to have anything to do with He. Based on what we already know, I encourage them to reject any contact with or overtures from He. The presidents of the National Academies and their foreign equivalents, the directors of the NIH and National Science Foundation and their foreign equivalents, while not pressuring scientists to avoid He, should make it clear that they approve the shunning of He, pending further light on the situation. Journals should take the same position. Funding agencies, particularly governmental ones, might have had a harder time ignoring applications before the official determination of He's guilt. It should now be easy in this case, but they should at least explore their legal powers to do so for collaborators of He who have not been convicted, or charged, as well as for similar future cases.

I do not recommend that the U.S. Academies, federal research funders, or foundations that support research perform their own investigation of He's actions. With the most important evidence in China, they just do not have the power to do so. (Although Rice—and perhaps others—need to investigate Michael Deem's role.) But these groups should be alert to information and judgments from China about the He experiment. China has already taken serious actions against He: his funding has been cancelled, he has been fired from his faculty position, and he has been convicted of crimes and sentenced to three years in prison. These Chinese criminal findings against him should be the basis for formal disqualification or other "blacklisting," at least as long as those determinations are credible. Perhaps, someday, a long life of repentance and good works by He might justify Science in readmitting him into its ranks—but not soon, both for his own sins and, more importantly, to "encourage the others."[6]

Snitching

The fact that many academics had hints, or direct knowledge, of what He was doing, but said nothing, raises harder issues. Each has said that he had conversations with He about human embryo gene editing. Almost everyone has said that he discouraged He from doing it. Several have said that they suspected he might be doing it anyway. A few have said they actually knew about the pregnancies some months before the babies were revealed. Not one of them disclosed his knowledge in advance, at all, to anyone.

I think they should have. But the word "snitching" conveys some of the difficulties of insisting on disclosure. Informing on others is sometimes socially required, but at the same time often socially repugnant. From siblings, to high school students, to employees, "informing the authorities" about a colleague's misbehavior will often get you labeled as a snitch. Or often as a "dirty snitch."

In addition to this basic social conditioning, science has strong conventions of confidentiality that allow colleagues to talk to each other without fear of being scooped. Both peer review, in publications, and review, in grant applications, typically include strong confidentiality requirements, such as the destruction of any paper or electronic copies of the submitted article or the grant application. That internal code of confidentiality presumably leads to more discussion and cross-fertilization of ideas—and better research. Destroying it could slow scientific progress.

More concretely, scientists who snitch will almost certainly ruin their relationship with the "snitched upon" colleague the same way that pediatricians who—in good faith and in response to strong state laws—report parents as potential child abusers may lose those parents and their children as patients. Informing scientists might even find themselves sued—successfully or not—for libel, slander, and various other torts. If the "snitch" is a competitor, as will often be the case, the legal claims might even be plausible. And the scientists who report may incur broader social costs from other colleagues and potential collaborators who shun them as snitches.

Against the backdrop of this social conditioning and the valuable conventions of confidentiality, should the scientists aware of He's activities have disclosed their conversations and suspicions? And, if so, to whom?

This is not a new question, to science or generally. It is not even new in discussions of the He affair: an editorial in *Science* magazine by the presidents of three national academies called for "an international mechanism that would enable scientists to raise concerns about cases of research that are not conforming to the accepted principles or standards."[7]

This question has arisen before in the biosciences with respect to so-called dual-use technologies, those that could be used for good purposes or for evil ones, such as biological warfare. It also comes up in more routine situations where someone is aware of wrongdoing and we as a society want to encourage or protect their whistleblowing. *Qui tam* statutes, giving whistleblowers some of the proceeds of suits against wrongdoers, date at least as far back as the Civil War. Their use has continued and expanded in recent years, especially in cases where fraud against the U.S. government is alleged.[8]

Although in English common law it was an offense called "misprision of felony" to fail to report crimes, no such general duty to report crimes is now widely recognized in the United States. (A federal statute, 18 U.S.C. § 4, appears to make failure to report a felony a crime, but the courts have held that it requires more than failure to report but active concealment of the crime.[9]) Sometimes, however, "snitching" on particular illegal activities is required. For example, all U.S. states have statutes requiring either specific professionals or broader groups to report elder abuse[10] and child abuse.[11] Failure of covered people to make these report is, in many places, a crime.

The He affair is simply the most recent example of why Science should think, hard, about encouraging, or even requiring, scientists to inform *someone* of their concerns about ongoing research. I am largely convinced that such an obligation should be created. But the details are important, and those are tricky to get right.

What would be the obligation? Only to disclose behavior you believe to be illegal, or is "unethical" enough? Is this a binding legal or ethical obligation or a guideline or aspiration? Do you have to be certain of the other's misbehavior, or have to have "clear and convincing evidence," or have to have a "preponderance of the evidence," or just have to have "reasonable suspicion"? What kinds of things should be reported? Plagiarism? Inappropriate authorship credit or order? Minor unapproved changes in a human subjects protocol? Dangerous work? Unethical work? Illegal work?

Then we hit the question of whom to tell. At least one of He Jiankui's confidants, Matthew Porteus, has said that he thought about telling someone about He's likely plans, but he did not know where to go.[12] This is a real problem, especially when the two scientists are not at the same institution. When they are at the same university, a word to the relevant ombudsperson, department chair, dean, research vice president, or president might do the trick. But how would, say, Stanford Professor Matt Porteus go about contacting someone at China's Southern University of Science and Technology? It is useless to tell people to gather up their courage and take action, unless you tell them where and how to report the misbehavior of colleagues.

We should create "scientific snitching" bodies. They could be located in academic institutions, in funding bodies, in national governments, or even in some kind of international organization. Scientists should be told they have a duty to report to this entity some kinds of illegal, unethical, or dangerous research. Congress, or another national legislature, could even give immunity from lawsuits for those who report when they act in good faith.

But we also should spare a moment's thought, and pity, for the people who receive these reports. Some will be from disgruntled coworkers, or jealous rivals, or from the apparently mentally ill. How much chaff will need to be sifted to reveal how little grain? And who in the world would want that job?

At this point I am not sure exactly how "snitching bodies" should work, but I am convinced that Science needs to think hard about encouraging internal reporting of dangerous, unethical, or illegal research. The alternative may well be ham-fisted external requirements, or yet more loss of trust in the beneficent motives and results of science—or both. We need further study and thought on the details of the idea. We can examine precedents, such as requirements for medical professionals to report their patients for abuse and colleagues for practicing while impaired. Academic honor codes provide other useful precedents. The National Academies, or some similar group, should convene a committee to study the feasibility of such a reporting requirement and, within a short time, report with recommendations on whether and how to make it happen.

Humility

Perhaps most importantly, Science needs to express—and to feel—humility. The He affair fed public concerns about mad, bad, and rogue scientists. After all, He Jiankui *was* a rogue scientist. He proceeded secretly, and without public discussion, to do something that he knew, or should have known, would be widely condemned. He has been convicted of committing fraud to do so. He's actions led many in the public to worry that scientists were pursuing their schemes with no regard for the law or for the opinions of their fellow citizens, citizens who were largely footing their bills. Science needs to make clear that it cannot, will not, and does not want to pursue research that is not acceptable to its society.

Before the He affair, scientists' statements about human genome editing openly acknowledged the importance of public opinion. A March 2015 article in *Science*, many of whose authors became members of the organizing committees of the international human genome editing summits, said science should

> Strongly discourage, even in those countries with lax jurisdictions where it might be permitted, any attempts at germline genome modification for clinical application in humans, while societal, environmental, and ethical implications of such activity are discussed among scientific and governmental organizations. . . .[13]

The Summary Statement of the organizing committee for the first International Summit, in December 2015, said,

> It would be irresponsible to proceed with any clinical use of germline editing unless and until (i) the relevant safety and efficacy issues have been resolved, based on appropriate understanding and

balancing of risks, potential benefits, and alternatives, and (ii) there is broad societal consensus about the appropriateness of the proposed application.[14]

The report issued on Valentine's Day, 2017, by the U.S. National Academies of Sciences and Medicine said, "With respect to heritable germline editing, broad participation and input by the public and ongoing reassessment of both health and societal benefits and risks are particularly critical conditions for approval of clinical trials."[15]

The United Kingdom's Nuffield Council on Bioethics issued a report in July 2018 that said,

> We recommend that before any move is made to amend UK legislation to permit heritable genome editing interventions, there should be sufficient opportunity for broad and inclusive societal debate.[16]

What all of these findings have in common is the need for public buy-in—at least acceptance if not full approval or consensus—before proceeding with human germline genome editing. At the Hong Kong Summit where He revealed his work, David Baltimore, chair of the organizing committee, initially struck the right note. Immediately after He's appearance, Baltimore said, forthrightly, "There has been a failure of self-regulation by the scientific community because of a lack of transparency."[17] And, indeed, the closing statement of the organizing committee reiterated Baltimore's condemnation of He's work.[18]

But there were disturbing off-notes, both in the closing statement and in individual statements from prominent scientists. The closing statement said,

> The organizing committee concludes that the scientific understanding and technical requirements for clinical practice remain too uncertain and the risks too great to permit clinical trials of germline editing at this time. Progress over the last three years and the discussions at

the current summit, however, suggest that it is time to define a rigorous, responsible translational pathway toward such trials. . . .

A translational pathway to germline editing will require adhering to widely accepted standards for clinical research, including criteria articulated in genome editing guidance documents published in the last three years.

Such a pathway will require establishing standards for preclinical evidence and accuracy of gene modification, assessment of competency for practitioners of clinical trials, enforceable standards of professional behavior, and strong partnerships with patients and patient advocacy groups.

The closing statement also called for "continued international discussion of potential benefits, risks, and oversight of this rapidly advancing technology" and for strong partnerships with patient advocacy groups and broad public dialogue. That's fine. It did not, however, say that a "broad societal consensus" would be necessary before starting clinical trials. And it did not say that, before such trials start, "there should be sufficient opportunity for broad and inclusive societal debate." This closing statement could easily be read as "There are a lot of technical things scientists need to figure out before this can be done. The public should have a chance to comment, but they will not make the decisions. We will."

This impression was abetted by unfortunate statements alluding to the inevitability of human germline editing. For example, George Daley, a member of the organizing committee, one of the major speakers, Dean of Harvard Medical School, and someone I like and respect, said this at the Summit:

I want to suggest that I do think it's time to move forward from the prospects of ethical permissibility to start outlining what an actual pathway for clinical translation looks like. What would be the regulatory standards that a group would be held to in order to bring this technology forward?[19]

In Daley's statement, those regulatory standards did not include a societal consensus, or even social acceptance. He took a few bows toward society, but one could quite easily hear in his comments that scientists should be the ones to figure out when, and how, this new technology will be used. *Science* subsequently quoted Daley as saying, "We have to aspire to some kind of a universal agreement amongst scientists and clinicians about what's permissible. . . . Those who violate those international norms are held out in stark relief."[20] This quotation does not invite the public to contribute to this "universal" agreement. Subsequently, he said in an interview that "we haven't yet engaged the public in ways that would make it [genetically altered babies] acceptable."[21] He doesn't say, "We haven't yet engaged the public to see whether it would be genetically acceptable." And I think that's a serious mistake.

My complaint is not that the organizing committee or Daley said something wrong, but that they didn't say something both right and important. They did not say, let alone trumpet, the crucial need for public acceptance before anyone should use genome editing technology to make babies. At a time when rogue scientists, or Science itself, is being blamed for ignoring the public, its high and mighty representatives should expressly say the following: "Science is part of society. The decision to use this technology belongs in part to scientists, but ultimately to societies."

A few have said that but not many. On November 26, the Monday after the He revelations and the day before the Summit started, Feng Zhang, a scientist on the short list for CRISPR inventors, quoted the organizing committee's statement at the end of the first International Summit that it would be irresponsible to proceed without "broad societal consensus" and expressed the hope that "this year's summit will serve as a forum for deeper conversations about the implications of this news and provide guidance on how we as a global society can best benefit from gene editing."[22] And Robin Lovell-Badge ended a February 2019 article about the affair by saying, "The decision of whether or not techniques such as genome editing in human embryos ever go into the clinic cannot be a decision made by scientists or clinicians alone; it has to be a decision that is made by all those concerned."[23]

The odd thing is, of course, that the need for at least society's acceptance is a truism. If a country makes the use of genomic editing technology illegal—as many have, including (effectively) the United States—then work cannot proceed there. But the He affair marked an especially important time for Science to say this, openly and clearly. The primacy of public acceptance should have been the first sentence of any reaction by scientific leaders to He's work. Instead, it was largely absent. And this, I fear, was a self-inflicted wound.

Wearing my law professor hat, and not my bioethics hat, I am confident that demanding social acceptance before using human germline genome editing to make babies is both legally and politically right. And Science would benefit if its spokespersons made it crystal clear that they accept—and in fact agree with—that demand. Science cannot exist, let alone thrive, without the continuing financial, legal, and political support of the societies in which it works. Its leaders need to say so—early, often, and loudly.

But what has Science done so far? It started slow but seems to be gathering momentum, particularly through an international commission created by the U.S. National Academies of Sciences and Medicine and the Royal Society; a published call from prominent scientists for a moratorium on human germline genome editing; and an advisory committee created by the WHO. At the same time, it has faced another scientist's announcement of his plans, or, at least, hopes, to do more germline genome editing.

On December 14, 2018, the presidents of the U.S. National Academies of Sciences and of Medicine and the Chinese Academy of Sciences published an editorial in *Science*, calling for several actions.[24] One is an increasingly detailed and concrete delineation of what conditions would need to be met to allow ethical human germline genome editing (a moratorium of a sort); another is a mechanism for scientists to report unethical research. For five and a half months after the revelation of the He experiment and five months after that editorial was published, nothing happened. Then, on May 22, 2019, the U.S. Academies and the U.K. Royal Society announced the formation of an International Commission on the Clinical Use of Human Germline Editing

> to develop a framework for scientists, clinicians, and regulatory authorities to consider when assessing potential clinical applications of human germline genome editing. The framework will identify a number of scientific, medical, and ethical requirements that should be considered, and could inform the development of a potential pathway from research to clinical use—if society concludes that heritable human genome editing applications are acceptable.[25]

I do applaud the end of the last sentence of the quotation—"if society concludes that heritable human genome editing applications are acceptable"—though I wish it had been said earlier, more frequently, and in ways that will get more attention than a press release announcing a new commission.

The commission has 18 members from 10 countries and has been asked to issue a final report by Spring 2020. Its first public meeting was in Washington, DC, on August 13, 2019; its second meeting was held in London on November 14 and 15, 2019. It also held four webinars during October 2019.[26] It is not clear how many private meetings the commission has had; given that its members come from 10 countries, I suspect many of its meetings will be through telephone or videoconferencing. Spring of 2020 seems to me an ambitious target for completion of their report— by the time you read this, we will know whether they made it.

On the same day in December 2018 as the publication of the *Science* editorial, the WHO stated it was going to establish a "global multi-disciplinary expert panel to examine the scientific, ethical, social and legal challenges associated with human genome editing (both somatic and germ cell). . . ."[27] On February 14, 2019, the WHO announced the 18 panel members, drawn from many countries and cochaired by Justice Edwin Cameron of the Republic of South Africa and the former commissioner of the FDA, Dr. Margaret Hamburg.[28] Some of the members are well-known to me, and I strongly endorse their selection. Others are unknown to me; one or two I know, but find surprising. The committee was asked to deliver its final report to the WHO director general within 18 months, so by August 2020.

Four weeks later, on March 13, 2019, a group of leading scientists, including many early CRISPR pioneers, released a statement in *Nature* calling for a formal "moratorium" on human germline gene editing.[29] Although they do not have formal leadership positions in "Science," they include some very important and well-respected scientists, such as Paul Berg, Nobel Prize winner and one of the parents of the Asilomar meeting on recombinant DNA; Emmanuelle Charpentier and Feng Zhang, two of the people viewed as among the most important inventors of CRISPR; Eric Lander, the director of the scientifically powerful

Broad Institute, a joint venture of Harvard and the Massachusetts Institute of Technology; and 14 others.

In the same issue of *Nature*, the presidents of three of the four groups that sponsored the International Genome Editing Summit in Hong Kong (the U.S. National Academies of Medicine and of Sciences and the U.K.'s Royal Society) published a four paragraph response, which stated that they shared the concerns of Berg's group and discussed what the Academies were doing.[30] It ended by saying,

> As emphasized previously by our academies and others, we also recognize that—beyond the scientific and medical communities—we must achieve broad societal consensus before making any decisions, given the global implications of heritable genome editing.

The presidents' response did not, however, endorse the call for a moratorium. In another letter in the same issue, NIH Director Francis Collins and Carrie Wolinetz, head of the NIH Science Policy office, expressed their strong support for the Berg group's call for a moratorium.[31]

Usually, I would view this dispute over whether or not to call for something termed a "moratorium" as an unhelpful kind of symbolic politics. A "moratorium" is defined as a "temporary prohibition of an activity."[32] (As the statement in *Nature* itself notes, "By 'global moratorium', we do not mean a permanent ban.") Likewise, the various ethics statements before the He announcement, discussed in chapter 4, said germline editing of babies should not then be done—in effect, a moratorium. Indeed, most countries where this work could easily be done already prohibit it, with bans that are not expressly temporary. When the work is already illegal in the United Kingdom, the United States, most of Europe, and (now) China, what does a call for a moratorium add?

The call in *Nature* for a moratorium threads the needle, in a way. Its authors seek an "international framework," rather than a "purely regulatory approach" or an "international treaty." The authors seem to recognize, at least implicitly, that a binding enforceable international agreement would be highly uncertain even after years of work. An "international framework," meanwhile, could be supported by a coordinating body under an existing organization like WHO, in which the coordinating body would "convene ongoing discussions and specific consultation once a nation announces publicly that it is considering permitting a particular application" of human germline genome editing.[33]

To me, some of the calls for a moratorium are at least partially political theater. "We oppose this more than you do; you resist using the word 'moratorium' so we will insist you use it so that we can win." I don't often like political theater. I prefer my politics and policies to be substantive.

But these moves, especially by the *Nature* authors, are also, in part, efforts by those who are frustrated that Science has not been clear enough about needing public acceptance to regain public trust. The authors of the *Nature* letter make a clear statement on the need for societal consensus:

> [C]linical germline editing should not proceed for any application without broad societal consensus on the appropriateness of altering a fundamental aspect of humanity for a particular purpose. Unless a wide range of voices are equitably engaged from the outset, efforts will lack legitimacy and might backfire.[34]

I understand and agree with the impulse for broader societal consensus; I just don't see the importance of the "M" word, except perhaps as a good tactical attempt at regaining public trust.

Meanwhile, six days after the *Nature* comment was published, and about a month after WHO announced the membership of its panel, the WHO panel took its first action. The Advisory Committee on Developing Global Standards for Governance and Oversight of Human Genome Editing, as it was now called, announced at the end of its first meeting on March 19 that it had recommended that WHO create a registry covering all studies of clinical applications for both somatic cell and germline genome editing.[35] It called for the registry to be transparent (and presumably public). The report on the meeting said a failure to register relevant research should be considered "a fundamental violation of responsible research." It urged that scientific publishers and funders of research require participation. (At least *Nature* indicated at least general support for the idea.[36]) The committee established a working group to design the registry in all its details.[37]

The meeting report said the committee "agreed with the views previously expressed that 'it would be irresponsible at this time for anyone to proceed with clinical applications of human germline genome editing.'"[38] In the virtual press conference after the meeting, Dr. Hamburg, the cochair, answered a question about a moratorium by saying,

> Certainly the issue of whether a moratorium might have a role to play will be part of those discussions, but what we are really trying to do is look at the broader picture and look at how there can be a framework for responsible stewardship. I don't think that a vague moratorium is the answer to what needs to be done.[39]

On August 29, 2019, the day after the end of the committee's second meeting, the WHO announced that it accepted the advisory committee's recommendation for a registry.[40] The registry is to be based on an existing WHO entity, the International Clinical Trials Registry Platform. At least initially, it will include both somatic and germline clinical trials. The committee asked all relevant initiatives to register any trials, but it has no power to force registration. According to the same press release, at the second meeting the Secretary General of WHO, Tedros Adhanom Ghebreyesus, said,

> Since our last meeting, some scientists have announced their wish to edit the genome of embryos and bring them to term. This illustrates how important our work is, and how urgent. Human germline genome editing poses unique and unprecedented ethical and technical challenges. I have accepted the interim recommendations of WHO's Expert Advisory Committee that regulatory authorities in all countries should not allow any further work in this area until its implications have been properly considered.

At this writing, nearly six months after it was announced, the proposed registry does not yet seem to be operating. It does seem to me a useful first step, though the details (particularly around mandatory information disclosure, to which researchers, or companies, might be strongly opposed) may prove sticky. But it is just a start.

Meanwhile, as the Ghebreyesus statement indicated, the world confronted another researcher who wanted to edit embryos to make babies. This one, at least, announced his plans in advance.

On June 10, *Nature* published an article revealing that Russian molecular biologist Denis Rebrikov wanted to follow He Jiankui's lead by editing embryos to produce children with nonfunctional *CCR5* genes.[41] Rebrikov runs a genetics laboratory at Russia's largest IVF clinic, the Kulakov National Medical Research Center for Obstetrics, Gynecology and Perinatology in Moscow, and is a "provost for research" at the Pirogov Russian National Research Medical University. Interestingly, and perhaps tellingly, like He Jiankui, he has no apparently training or background in reproduction but works on molecular biology methods and the analysis of genome structure and function.[42]

Rebrikov told *Nature* that he hoped to transfer edited embryos for possible implantation by the end of 2019. Rebrikov's plan did differ from He's in three respects. First, he was planning to help couples where the mother, not the father, was HIV positive and responded poorly to anti-HIV drugs, which would increase, somewhat, the risk that a child would be infected before birth. Second, he talked publicly about his plans in advance. And, third, he made it clear that he was going to seek permission from Russian authorities for his work. Russia, although a member of the Council of Europe, is not a signatory to the Oviedo Protocol, so it is not bound by its prohibition on germline modification. According to the article, Russia does prohibit genetic engineering "in most circumstances, but it is unclear whether or how the rules would be enforced in relation to gene editing in an embryo." Rebrikov anticipated that the Russian Ministry of Health would clarify those rules soon.[43]

Within a month Rebrikov changed his planned target from *CCR5* to the *GJB2* gene.[44] Some variations in this gene cause deafness.[45] The genetic condition is inherited in an autosomal recessive manner—a person must have two "bad" copies of the gene to be deaf—but Rebrikov said that there were cases in Western Siberia of some couples who both had the same genetic condition. These people have only deafness-causing variations to pass on to any children, so genome editing would be the only way for them to have hearing children from their genes. Rebrikov said that he had already found five couples where both parents had this condition and did not want their children to be deaf.[46]

Although there had been some speculation that Russian President Vladimir Putin might think favorably of human germline genome editing,[47] in early October the Russian Ministry of Health announced that it fully supported the WHO position that current use would be inappropriate.[48] At about the same time Rebrikov told *Nature* that he was editing the *GJB2* gene in normal eggs, looking for off-target problems, but that he would not try to edit embryos to produce hearing babies without government approval. His plans to apply for such approval in October had been pushed back indefinitely.[49] And on November 26, 2019 (the anniversary, at least in the Eastern Hemisphere, of the unveiling of He's work), 14 Russian scientists published a short letter in *Nature*, saying,

> The Russian community of geneticists, clinicians and bioethicists have reached a consensus on the use of genome-editing technologies on human embryos and germ cells for clinical purposes. . . . They consider that such experiments are premature at this point.[50]

On November 8, the WHO committee posted a two-page document online, entitled "As We Explore Options for Global Governance, Caution Must Be Our Watchword,"[51] responding, in part, to Rebrikov's plans and the Russian Ministry of Health's rejection of them. It noted that "most importantly, the Ministry of Health's press release explicitly stated that the WHO position, 'supported by the Russian Federation, should be decisive in the formation of country policies in this area.'"

The brief statement went on to say that the WHO advisory committee was looking not for a single mechanism but for a governance structure that could do the following:

(i) Identify relevant issues, a range of specific mechanisms to address them, and be developed in collaboration with the widest possible range of institutions, organizations and peoples.

(ii) Be scalable, sustainable and appropriate for use at the international, regional, national and local levels.

(iii) Work in parts of the world where there is traditionally weaker regulation of scientific and clinical research and practice, and

where genome editing may not yet be pursued or invested in
with great intensity.

(iv) Provide those with specific governance roles for human genome
editing with the tools and guidance they need.

It announced that its current plans call for a governance framework guided by five principles: Transparency, inclusivity, responsible stewardship of science, fairness, and social justice.

At this point, at least, the Rebrikov initiative appears to have
stalled.[52] That he proposed it is some evidence that the response
to He Jiankui has not deterred all researchers from these thoughts.
On the other hand, that he announced it publicly and stated he
would not proceed without regulatory approval may well show
that the He affair has had some effects. And his announcement
may have spurred helpful steps by the WHO and the WHO committee. WHO is a notoriously bureaucratic and political organization, but, so far, it seems ahead in actively addressing the
aftermath for Science of He Jiankui.

The response of Science so far to the He Jiankui experiment
has been limited but promising. There has been insistence that
further experiments on babies are not now appropriate, statement of the importance of some role for public opinion, and the
announced formation of a registry. But Science has done nothing toward enforcing deterrence or creating "snitching" mechanisms, and not enough, I think, to express humility. It should.

Chinese Responses

The response in China has been quite different, and much more
promising.

It is not a shock that the He experiment took place in China.
China has poured vast sums into biological, and particularly

genetic, research in the past two decades, and this investment has paid off. It is now clearly one of the two or three most important countries in the world for genetics research, still behind the United States, but challenging (or perhaps surpassing) the United Kingdom. With the funding has come a "Wild West" atmosphere, where Chinese scientists talk optimistically about understanding the genetic roots of human intelligence[53] and blithely use genome editing (TALENs in this case) to miniaturize pigs;[54] and use CRISPR to add longer hair and more meat to domestic goats;[55] to add muscle to beagles;[56] to modify the DNA of monkeys[57] and of human embryos;[58] and, ultimately, to edit the germline genome of human babies. And they do so with pride in the great advances of Chinese science, with a degree of nationalism that seems to me about the same as that of the United States.

Some voices in the West want to pick fights with China, for various reasons. There's nothing like a dangerous and worrisome international rival to spark increased domestic interest and funding. But one of the arguments advanced, more or less subtly, is an ultimately racist one that the Chinese have no ethics—they are "the Yellow Peril," unassimilable, incomprehensible, inevitably "the other."

As a Californian, I am painfully aware of how my state—and indeed, the founders of Stanford University, my employer and undergraduate school—used and abused imported Chinese laborers. And then California (though not the Stanfords), in a burst of populist racism, turned against them. Californians lobbied successfully for a ban on Chinese immigration to the United States in the 1875 Page Act, banning immigration of Chinese women, and the 1882 Chinese Exclusion Act, banning immigration of Chinese laborers.[59] In addition to federal law, California

and its counties and cities adopted further stringent and oner-
ous restrictions on the Chinese who were already in America, as
immigrants or as native-born citizens.[60] I find this a shameful
memory.

But sometimes it feels like more than a memory, as I see people decrying the lack of ethics by the Chinese. Of course, on
some points, the Chinese government will, of course, act unethically (as will, I suspect, all governments) and, on many points,
China really is a rival of the United States for various kinds of
regional and world leadership. More pertinent to this discussion:
China's treatment of HIV-positive people is deeply disturbing.[61]

But when it comes to research ethics, as far as I can tell the Chinese government and research establishment have roughly the
same sets of rules as the rest of the developed world, including
(for the most part) the United States. China is not as concerned
about research with human embryos as the United States and
some other Western countries are; neither are they as concerned
about genetic modifications to animals that do not clearly impli-
cate the modified animals' welfare. But, for both human subjects
research and most nonhuman research, both weighing of the
risks to the research subjects against the possible benefits and
(for humans) requirements for informed consent exist in China.

I think the differences have been more in implementation
than in ethical principles or rules. Some of the Chinese reports,
particularly the Xinhua report of the Guangdong investigation,
cast He Jiankui as a rogue or lone wolf, acting alone and hard to
detect, deter, or stop. I think the real problem is that China did
not have the depth of regulations and regulatory bodies—and
regulatory mindset—necessary to stop him. The United States
has a big and often annoyingly bureaucratic set of structures that
control most research, among them IRBs, Institutional Animal

Care and Use Committees, Biosafety Committees, and Embryonic Stem Cell Research Oversight Committees. Universities, research institutes, and pharmaceutical and biotech companies typically have specific research oversight departments to manage them. The local committees oversee compliance with the rules, but they also report to the federal government. The government ultimately wields the power to bar federal funding at noncompliant institutions or to disqualify some research from FDA consideration, power that is taken seriously.

Some institutions have made a great effort to streamline their systems, but I doubt that any institution has a fast, easy, or "fun" system. Scientists almost everywhere will complain about the delays they go through to cross every "i" and dot every "t" in their human subjects or animal use protocols. But the system does constrain behavior.

China comes from a very different tradition, at least recently. Although imperial China may have been the world's longest surviving bureaucracy (only the Catholic Church gives it a run for its money), the Communist Revolution produced an interregnum. For roughly 30 years between the 1949 victory of the revolution and the rise to power in 1979 of Deng Xiaoping, China was, in effect, a land without law. It had plenty of authority—as far too many people found out to their death and dismay—but that authority was exercised through often ad hoc decisions by the Communist Party and its leadership, not by legislative statutes implemented transparently and evenhandedly by bureaucracies.

For the last 40 years, China has been trying to "legalize" itself, passing statutes and creating bodies to implement them. This is particularly hard in a country roughly the geographical area of the United States, Brazil, Canada, or Australia, but with a population (respectively) roughly 4 times, 6 times, 30 times, and 55 times as large. It seems not to be as "law-focused" as a developed Western society, and may never be, but it is moving in that direction even if not evenly. Threats to the Communist Party's power are carefully watched and responded to, through legal and nonlegal mechanisms. Control over biomedical research is not nearly as great a priority.

China requires approval of human subjects research by local committees. It has regulations concerning ethical and unethical genome research. But, prior to the He affair, it did not have very powerful structures to implement those aspirations, or much experience of taking action to enforce them. I am writing this a year after He's experiment was revealed. International science has begun to respond, but China has already taken several important steps.

First, it took some rapid actions in response to the He revelations. In less than a week after the announcement of He's work, China suspended all of his research.[62] In December 2018, the Chinese Education Ministry "sent notices to universities requiring self-checks on research related to gene editing," telling the AP by email that "it called on educational institutions to strengthen management of scientific research ethics and inspect research involving gene-editing technology."[63]

Then, in January 2019, China's government announced its view that He's work was illegal as violating a provision in an "annex of a 2003 ministerial guidance to IVF clinics" banning genetic editing for reproduction.[64] The directive is not entirely clear, and, to this American lawyer's eyes, whether it would invoke any kind of sanctions, criminal or civil, was not clear, but legal systems differ. Based on the results of He's trials, it appears the court accepted retrospectively the official interpretation of the regulation, although the uncertainty about the exact nature of the charges against He leaves some uncertainty. He's conviction, and those of his co-defendants, and the now-clear official interpretation of the annex, should, in itself, deter this kind of research in one of the world's three leading genomics countries, one that is the home of about 15 percent of humanity.

Later that January, China's president, Xi Jinping, called for tighter regulation of gene editing through new legislation.[65] On February 27, 2019, China announced new regulations on "high risk" technologies, including gene editing, which would be governed by the health department of China's cabinet, the State Council.[66] In early March, Jane Qiu reported in *STAT* that China had decided to set up

> [a] powerful new national medical ethics committee, which will approve all clinical trials involving high-risk biomedical technologies, [and which] is at the center of a regulatory shakeup Chinese authorities are planning in the aftermath of the widely condemned "CRISPR babies" experiment. . . .
>
> The technologies that will be regulated by the ethics committee are often new and are deemed risky either because of safety or moral concerns. They will include not only gene editing, but also cloning, cell therapy, xenotransplantation, mitochondrial replacement, and nanotechnology.[67]

With a planned 30-member staff, the national ethics committee—personally approved by President Xi Jinping—will report to the health committee of the State Council, China's cabinet, and have regulatory jurisdiction over nearly a dozen ministry-level agencies that fund or regulate medical research and applications.

Qiu quotes David Archard, chair of the United Kingdom's Nuffield Council, and Wisconsin law professor Alta Charo, cochair of the 2017 National Academies report, as pointing out successes with national oversight committees, but noting the importance of inspection powers, which are not clearly present in this proposal:

> What's lacking, said both Archard and Charo, is an inspection body that would work in conjunction with the national ethics committee. In both the U.S. and the U.K., inspections—often in the form of surprise visits—is [sic] a critical aspect of regulatory oversight. "It has turned out to be incredibly important because the reports that are delivered [to federal agencies] are often not quite correct or not quite complete," said Charo. "It's very important for Beijing to know what's going on at the local level."

And then, on May 20, 2019, China announced that the latest draft of its revised Civil Code would include "human genes and embryos in a section on personality rights to be protected" and "[e]xperiments on genes in adults or embryos that endanger human health or violate ethical norms can accordingly be seen as a violation of a person's fundamental rights."[68] This version of the revised Civil Code, which has been undergoing revisions since 2002, was finally adopted at the end of May 2020;[69] the human genome provisions were added to draft in the last year of its development.

It is not just Westerners who are calling for more action by China. Recall that 122 Chinese scientists and ethicists signed a letter decrying the He experiment, which was disseminated on WeChat less than 24 hours after the story broke. On May 8, 2019, four Chinese ethicists and researchers, Ruipeng Lei, Xiaomei Zhai, Wei Zhu, and Renzong Qiu, called in *Nature* for a "reboot" of ethics governance in China.[70] They provide six specific recommendations, summarized as regulate, register, monitor, inform, educate, and end discrimination. They also call for a much larger and more transparent international investigation into the He affair than the preliminary investigation

by Guangdong Province, the details of which, itself, have only
been made public through a short article from the national press
agency. According to an interview with Dr. Qiu, "Nothing suggests another investigation is underway."[71]

Their *Nature* article ends with this reminder, and hope:

> It has been only around 30 years since bioethics was established in
> China. And it is worth remembering that unethical research practices
> were rife in the West in the early days of ethics governance. Take
> the infamous Tuskegee study, in which the US Public Health Service
> tracked—but did not treat—399 black men with syphilis from 1932
> to 1972. Just as the revelation of that research prompted the 1978
> Belmont Report, which protects human participants in studies or
> clinical trials, the 'CRISPR babies' scandal must catalyse an overhaul
> of science and ethics governance in China.[72]

When I began writing this book, this section was going to
call on China to act quickly to bring this research under greater
regulatory control. China largely has beaten me to it; I can only
call for China to be serious and diligent in implementing its pro-
posals, and the extensions suggested in the bioethicists' article.
And for us all to have a moment of deeply ambivalent reflection
on the ability of some governments to move quickly while other
governments seem to be unable to move at all.

IV Human Germline Genome Editing Generally—Now What?

11   Is Human Germline Genome Editing Inherently Bad?

No. At least, I don't think so.

More than 15 years ago, I reviewed two books about making changes in a human's DNA that would get into the eggs or sperm and so could be passed down to future generations.[1] Even then, the debate had been raging for decades, but it remained largely theoretical until November 25, 2019, when the story broke that He Jiankui had used CRISPR to edit the DNA of human embryos, two of which, by sometime that October, had become baby girls.

You have already read, at length, my views on just how bad the He experiment was. (Very bad.) A bit of my analysis hinges on the fact that what He did was human heritable genome editing and hence violated an international scientific consensus, but the rest of it does not. If He's experiment had been on blood-forming stem cells in newborns—nonheritable somatic cell gene editing—his experiment still should have been condemned.

But what *about* human germline genome editing? Outside the context of He Jiankui's work, is it bad? Is it good? Or, more plausibly, under what circumstances is it *how* bad versus *how* good? The next three chapters discuss this. This chapter asks whether there is something about human germline genome editing that makes it inherently, always, and in all circumstances, bad. The next chapter looks at arguments that say it might be bad in some circumstances. And the third asks, "Just what is it good for—and how good would it be?"

My views on the first point probably do not raise any particularly new issues from the last 40 years of discussions, but they may have some added value because they are informed by the context of the He Jiankui experiment. The most basic argument against human germline gene editing is that it is, inherently, intrinsically, and in all circumstances, morally wrong. By editing the human germline genome, we presume to do what humans should not attempt. I reject that view, as do many, but not all, of those who have recently examined this issue. Before getting to my critique, let's briefly survey three major ethics reports on the topic, each of which, in different ways and answering different arguments, also rejects the conclusion that human germline genome editing is inherently unethical.[2]

The Three Reports

The last three years have seen the publication of three reports, from three different countries, that have focused entirely or substantially on reproductive uses of human germline genome editing. Each contained express conclusions and recommendations about the technology. You have heard about two of them before—the first, published on Valentine's Day, 2017, by the U.S. National Academies,[3] and the second, issued in July 2018, by the United Kingdom's Nuffield Council on Bioethics.[4] The most recent, published in May 2019, after the revelation of the He Jiankui experiment, is a report from the German Ethics Council.[5] Each report merits careful reading, in part because although not one of them concludes that human germline genome editing is inherently unethical, each report gets to this conclusion some-what differently. I have already discussed many aspects of the first two reports in chapter 4; here I will hit only the most relevant general points before moving on to the German report.

The National Academies' Valentine's Day report has eight chapters addressing five main substantive topics: basic research using genome editing, somatic genome editing, heritable genomic editing, enhancement, and public engagement, and detailed recommendations. The report found no convincing reason to reject all efforts at what it called "heritable genomic editing." It concluded,

> If the technical challenges are overcome and potential benefits were reasonable in light of the risks, clinical trials could be initiated, limited to the most compelling circumstances, and subject to a comprehensive oversight framework that would protect the research subjects and their descendants; and have sufficient safeguards in place to protect against inappropriate expansion to uses that are less compelling or less well understood.[6]

On its way to the conclusion that, in some circumstances, human germline genome editing *might* be ethical, the committee surveyed, and rejected "concerns about diminishing the dignity of humans and respect for their variety; failing to appreciate the importance of the natural world; and a lack of humility about our wisdom and powers of control when altering that world or the people within it . . ."[7]

The Valentine's Day report comes closest to my analysis in this book in its discussion of "A 'Natural' Human Genome and the Appropriate Degree of Human Intervention."[8] It then discusses Human Dignity and the Fear of Eugenics, Economic and Social Justice, and the Slippery Slope.[9] The report ultimately lays out, in its recommendation 5–1, the 10 conditions, discussed above in chapters 4 and 9, that would have to be met before a "[c]linical trial using heritable HGE [human genome editing] should be permitted."

The July 2018 Nuffield Council on Bioethics report, *Genome Editing and Human Reproduction*, following a September 2016 report, *Genome Editing: An Ethical Review*,[10] which was an effort to map the broad landscape of the questions presented by the technology, concluded

> that interventions of this kind to influence the characteristics of future generations could be ethically acceptable, provided if, and only if, two principles are satisfied: first, that such interventions are intended to secure, and are consistent with, the welfare of a person who may be born as a consequence, and second, that any such interventions would uphold principles of social justice and solidarity—by this we mean that such interventions should not produce or exacerbate social division, or marginalise or disadvantage groups in society.

The report states that the prospects of ethically permissible human germline genome editing, and the changes to U.K. law needed to permit it, are remote and may never appear. But the council also notes that broad social debate and discussions leading to an international framework governing genome editing need to happen sooner rather than later.

The second chapter of the five-chapter Nuffield Council report is particularly interesting for its attention to social setting and to the possibilities of technological momentum, function creep, slippery slopes, and possible evolution of normative values. Its depth of analysis on these issues is, in my experience, unusual—and unusually interesting. The third chapter focuses on ethical issues, with subtle and thorough discussions, leading to the conclusion that "none of the considerations raised yields an ethical principle that would constitute a categorical reason to prohibit heritable genome editing interventions." The fourth chapter, on governance, considers what governance changes may be necessary, in the United Kingdom and in the international order ("the problem of geo-ethics"), to deal appropriately with this technology. For the United Kingdom it recommends that the use of these techniques

> should not be permitted until risks of adverse outcomes have been thoroughly assessed and then only on a case-by-case basis, licensed and regulated under the system currently overseen by the HFEA, and within the context of a carefully monitored study, with comprehensive follow-up arrangements in place.

The most recent report, from Germany, was published in May 2019. In September 2017 the *Deutscher Ethikrat* (the German Ethics Council) had called for a broad social debate on human germline interventions. On May 9, 2019, the council issued *Eingriffe in die menschliche Keimbahn*, a report focused exclusively on human germline uses of genome editing.[11] The Executive Summary and Recommendations have been published in English translation as *Intervening in the Human Germline*.[12] The following discussion is based on that translation.

Because of its Nazi history, postwar Germany has been very wary of human genetic interventions. (Recall that Germany has not signed the Oviedo Convention because it found it too permissive, unlike the United Kingdom, which found it too restrictive.) Since 1990, German laws have criminalized germline modification for reproductive purposes.[13] The report took as its mission the need to determine, in light of the prospect of easier and more accurate germline intervention, "the question whether the previous categorical rejection of germline interventions can be upheld or whether it must undergo a new ethical assessment."[14]

The report exhaustively assesses various ethical arguments relevant to human germline genome interventions. It rejects a solely consequentialist ethical analysis, instead stressing the importance of "human dignity": "that value which is resistant to any trade-offs and which is due to man as such and independently of all social provisions: man is regarded as an 'end in himself.'" It discusses, though largely dismisses, the idea of the "dignity of the human species." It also talks about "the ethical concept of naturalness," which it finds based both on a belief that the human germline is the "symbolic heritage of mankind" and on the fear that germline interventions will necessarily have unanticipated and uncontrollable effects. But the report also notes that "the argument of naturalness is often used as a placeholder to articulate a vague unease about the mechanisation of the world." Other concepts discussed include beneficence, non-maleficence, justice, solidarity, and freedom.

To simplify the report's nuanced discussion, it sees no categorical barriers for any plausible applications of human germline genome editing but finds more troubling enhancement and reduction of risks of non-Mendelian diseases. As to Mendelian diseases, the report uses the example of a couple who wish to have a genetic child, but where each prospective parent has cystic fibrosis, making PGD unavailing. It concludes,

> Overall, for the ethical assessment of germline interventions in monogenic hereditary diseases (assuming sufficient safety and efficacy of the technology), no categorical reasons for prohibiting such interventions can be derived from the application of the ethical concepts. Rather, the ethical concepts of the protection of life, of freedom and of beneficence suggest for some a duty to permit such interventions. Against this backdrop, considerations of non-maleficence, justice and solidarity do not provide any substantial arguments against the interventions.[15]

My Perspective

I appreciate the breadth of the arguments from those three recent national reports—and will address some of them in the next chapter—but my argument in this chapter is aimed at what I believe is the core premise driving many of those who oppose the technology: the unique importance of the human germline genome. Opponents urge that "The Human Germline Genome" is the common property of all humanity, the essence of our species, and thus must be kept inviolate. Or, more specifically, as UNESCO stated in a formal declaration, "The human genome underlies the fundamental unity of all members of the human family, as well as the recognition of their inherent dignity and diversity. In a symbolic sense, it is the heritage of humanity."[16] It seems the equivalent of the Ark of the Covenant—and, as anyone who saw *Raiders of the Lost Ark* knows, it cannot be allowed to fall into the wrong hands.

But *none* of this is true. There *is* no "Human Germline Genome," and, to the extent there is, it has changed, is changing, and will continue to change inevitably from generation to generation, often as a result of human actions. And, in any event, not all CRISPR-induced changes should necessarily count as "changes."

As I've had you read about human germline genome editing for many long pages now, how can I say there isn't a human germline genome? It is not that there isn't one—there are, in fact, currently about 7.5 billion human germline genomes. Every living person has a "germline genome," and each one is different. Even identical twins have slightly different germline genomes due to mutations in their dividing cells, some of which became eggs or sperm, after the time they split from one early embryo. Even within one person, there may be multiple germ-line genomes. If one of the precursors to a person's eggs or sperm had a mutation in one of two daughter cells it divided into, the germ cells descended from those two daughters will—or may, depending on the chances of meiosis—have a different genome from egg or sperm cells from the other daughter cell.

Whose genome is *the* human germline genome? Well, mine, of course—though you may have a different view. But even then, which genome found in my sperm precursor cells forms the human germline genome?

There is *a* human germline genome in any one of the precursors to any one egg or sperm of one person, and it can defined quite precisely as an exact sequence of about 6.4 billion As, Cs, Gs, and Ts. (At least, exact within the small but not trivial measurement errors in our current sequencing technology.) But if you compare human germline genomes across different people, the picture is very different. You get a very fuzzy ball. Counting only the roughly 23,000 protein-coding genes (a few percent, though a few very important percent, of the whole genome), it is a fuzzy ball in 23,000-dimensional space. For any particular gene, perhaps 10 percent of people will have exactly the same, most common sequence, but the other 90 percent will have thousands of slightly different variations. Each variation is in a slightly different spot for that one gene, and that degree of variation is repeated for each of the other 22,999 genes.

The genome of each one of us has a particular location for each of our genes, but each gene has slightly different locations in most people. The result is a very fuzzy ball of dots. It hovers around humans, but much of it—some of the genetic variations in the protein-coding genes—are found in other primates, other mammals, other vertebrates, and other life-forms, including bacteria. If the sequence of one of your homeobox genes (a very basic type of gene that produces proteins that help determine the structure of multicellular animals) is the same as its sequence in (most) fruit flies, is it part of "the human germline genome"? Is it part of "The Fruit Fly Germline Genome"? Is it part of both?

That's before even thinking about the possibility of inherited epigenetic changes. This is a trendy, but very uncertain, idea that sometimes changes not in DNA sequence but in how it is expressed get passed from parent to child. There is evidence for this in a few cases in laboratory mice; there is a (very) little suggestive evidence for it in humans. If these exist (I am currently agnostic but leaning skeptical), they would add further variation and complexity to "the human germline genome."[17]

Having said that, there will be parts of the 23,000-dimensional ball that are found in (almost) all humans and not found (much) in nonhumans. It may not be one, infallible text, but perhaps a braided stream of texts that are sufficiently different from others to deserve a name. But it is not a simple "thing" (let alone the essence of humanity), and it may be hard to define it in a way that excludes a gene-edited baby from "the human germline genome." Fundamentally, I hope this has led you to question—sharply—the very concept of "the human germline genome."

But, even if you continue to believe that there is such a "thing" with determinate boundaries, that just brings us to the verb, "to change." Human germline genome editing would change *a* human germline genome—that of the edited embryo—but would it change *the* human germline genome, whatever that is, in any ethically important way?

Note, first, that the human germline genome is *always* changing. Every baby has a different germline genome than any human has ever had before (even identical twins as a result of post-separation mutations, if only slightly). That's what sex does—it merges parental variants into never-before-seen new combinations of variations. Accepting that each person's genome is part of "the human germline genome," each person adds new points to it, thus changing it.

Second, every baby's germline genome has been changed in other ways. Thanks to mutations, both in the parents' germlines and in the baby's early embryo and germline, it is not made up of all the exact variants that the baby's parents inherited from their parents. And its children, if any, will not inherit exactly the same variations it inherited.

But it gets even more complicated. Not only does the human germline genome change over time but humans indirectly "cause" those changes. Widespread adoption of agriculture changed the human germline genome. We have more starch digesting genes than our hunter-gatherer ancestors, not because some early farmer inserted them but because, given the diet available, babies who could digest starch better thrived and had more (better starch-digesting) children, and so on.[18] Natural selection acts on humans, even when the "nature" has been modified by ourselves. When humans change the environment in ways relevant to themselves, and thereby change the fitness pressures that environment imposes, humans inevitably change which genetic variations get selected for and against. (This is not the only cause of changes in human genomes over time and may not be the most important; random variations called "genetic drift" play a very big role, but they are not chargeable to any human action.)

Medicine provides great examples. By allowing people with illnesses that have a genetic component to have greater reproductive success (to live longer, to have more children, and to have more healthy children), doctors soften natural selection's action against disease-causing or disease-predisposing genetic variants. Type 1 diabetes (formerly known as juvenile onset diabetes) is a great example. Before Banting and Best discovered how to inject insulin to keep type 1 diabetics alive in the 1920s, most of them died before they had a chance to pass their predisposing genetic variations on. So, the last 100 years should have seen an increase in the incidence of such "high-risk" variations.[19]

Apart from this constant flux in the human germline genome, often influenced by human actions, it isn't clear whether the kinds of editing most often proposed should even count as changes. Most proposed uses for human germline genome editing are to prevent disease. A dangerous, almost always rare, and never common, genetic variation is sought to be replaced by the safe, most common version of that gene. A *huntingtin* gene with, say, 58 cytosine-adenine-guanine (CAG) repeats—leading invariably to Huntington's disease—is edited into one with fewer than 37 such repeats—and complete protection from the disease. Roughly 49,999 humans out of 50,000 have a "safe" version of the *huntingtin* gene. Is it "changing" *the* human germline genome to change that one in 50,000 person back to the overwhelming species-wide norm? It changes that person's germline genome, and it changes, *very* slightly, the ratio of the different versions of this gene, but only by increasing the number of "normal" versions.

This turning of disease-causing variations into safe ones will be the early (though not, as it turned out, the first) target for human germline genome editing. But some people, including, in theory, George Church, and, in practice, He Jiankui, argue for modifying common variants into much less common ones, variants that seem to be protective. But that raises a variation of the same question. Is it changing the human germline genome to change the variant frequency by changing the most common variants into those known to exist but to be uncommon? Would it change the human germline genome to change some people's genes to make their blood group Rh negative, currently about 6 percent of the world's population, instead of Rh positive, 94 percent?[20]

There is another possible step—editing human genes into variations never seen before in humans, sometimes seen in other species and in other cases created by scientists, from scratch or (more likely) by making plausibly improving changes in the gene (and the resulting protein). Only a few people are talking seriously about this possibility, which even I would call "changing" the (fuzzy) human germline genome. Even with my concession, two question arise. First, we cannot be sure that the variant is not, or has never before been, seen in humans. How can we know that not one of the 7.5 billion of us—or the scores of billions of humans who came before us—already carries that variation as a result of a "natural" (or, at least, not engineered) mutation?

Second, and more important, even if it had never been seen in the human germline genome before, would that make adding it inherently unethical? I have argued that the human germline genome is not a sacred essence of humanity, but barely a thing at all, that it is constantly changing, that human actions have long changed it, and that changes that just modify the prevalence of different variations already found in humans might not really be changes to the abstract concept. Legitimate questions should be asked about particular uses of human germline genome editing, such as questions about safety, equity, and enhancement. But if an introduced change is appropriate in light of those questions, does anything about the nature of germline editing nature make it ethically questionable?

One might argue that human germline genome changes are irreversible, or less reversible, than some other interventions. But is that true? A mistaken genome change could presumably be reedited, in a living person or in that person's germline (or embryos), to reverse the error. Or it could be selected against in the individual's offspring, through PGD or otherwise. It could be too late to avoid harm to the edited person, but that mistake will not have to pass on from generation to generation.

Human germline genome editing does raise important questions about safety, coercion, equity, diversity, and enhancement, which are addressed in the next chapter. These are not unique to editing the human germline genome. They also apply to somatic cell DNA editing, to new drugs, to smartphones, to climate change, and to many other changes from technologies. Questions about reversibility also apply; the social effects of cell phones are probably less reversible than genome edits. The fact that a technological change is in "the human germline genome" should have no special ethical weight.

12 Could Human Germline Genome Editing Sometimes Be Bad?

Of course. What technology can't be?

And that is particularly true of technologies that strike as close to our humanity as those that involve reproduction. Human germline genome editing could certainly be used in ways that cause needless harm or more harm than good. Important questions are whether those results can—and likely will—be averted? This chapter surveys a few of the very real issues that are most discussed as problems for this technology (and many others): safety, coercion, equity, diversity, and enhancement. But first I want to dispose of a common, but illogical, objection to all of human germline genome editing: "unnaturalness."[1]

Some will argue against human germline genome editing on the ground that it is unnatural (the secular version) or against God's will (the religious version). Both versions are likely to have substantial visceral and hence political force. Neither has much logical support. The biggest problem for those who value naturalness is that almost nothing in our lives—including the lives of almost all the naturalness proponents—is natural. Our clothes are not natural fur, our diets (even if rigorously, albeit foolishly, consisting of non–genetically modified organisms) are not plucked from natural sources unaffected by human interference. Most notably our medicine is, by its very nature, "unnatural." It seeks to preserve life and health that "Nature" wants to take away.

The religious version has a similar problem. Except in rare cases in which a holy scripture is very express—the prohibition of boiling a baby goat in its mother's milk appears twice in Exodus and once in Deuteronomy[2]—some kind of interpretative process will be required to divine what God wants. And I am confident the Torah, the Bible, and the Koran have no express prohibition on human germline genome editing. (And probably not the Hindu holy texts either.)

Similarly, Judeo-Christian traditions hold that God created man (and woman) in his (and her?) own image. Altering humankind, thus modeled on the divine, might then be viewed as sacrilege, but we already "change" humans in myriad ways, from vaccinations to orthodontic braces to eyeglasses to education (literacy literally changes a part of the brain known as the letterbox region—what it responds to and what brain regions it connects to).

So one is left to try to distinguish, without clear textual support, between permissible deviations from nature (or the conditions of life when the various religious traditions were formed) and impermissible ones. It is hard to find any clear distinction between human germline genome editing and other modern innovations. Perhaps one could say that it affects future generations, but, of course, everything parents do with or to their children has that potential. Or one could single out reproduction as an area where change is taboo. But only a few people (outside of celibate priests in the Vatican) are willing to give up contraception or assisted reproduction. And very few indeed are willing to give up cesarean sections when necessary, anesthesia during labor and delivery, or, when still useful, forceps.

Don't get me wrong. If, let's say, an Amish person believes that God is fine with him using power tools but not with him driving, I have no objection to his living those beliefs. I do object to his imposing those views on others who do not hold them. Novel technologies deserve close attention because we may not immediately recognize harm they may cause, but it is the potential harm that is important for policy, not the novelty in itself. And so let's turn to some concrete potential harms that could arise from human germline genome editing.

Safety is a very real issue, especially as the people whose lives and health are put at risk are future babies who had no say in the decision to have their germline genomes edited. Safety is, of course, a crucial issue with all medical innovations (some of which will also be intended to be used for embryos, fetuses, or infants). We cope with it, with what seems to be pretty good success, with drugs and medical devices. Chapter 14 suggests how the safety of human germline genome editing could be tested and thus regulated. It does not, however, offer any way to guarantee safety absolutely.

Can and will "reasonable" safety be ensured? I do think regulatory systems like that of the FDA in the United States and the (very different) HFEA in the United Kingdom, and, no doubt, similar systems in many other countries and regions, should be able to do an adequate job, at least in determining whether the method seems safe enough to be clinically available. Ensuring continued safety once human germline genome editing is available may prove more challenging, but not insuperable.

The biggest problem is likely to come from "end runs" around these regulatory systems. Desperate people will likely find providers who will, for a high enough fee, provide unproven germline editing, often in foreign havens. The stem cell field has had a continuing struggle with "clinics" offering stem cell "cures" with no good evidence of either safety or efficacy. (We have seen, in passing, such an attempted end run in chapter 8 with John Zhang and mitochondrial transfer.) The same could happen here, although, given the alternatives to germline editing laid out in the next chapter, there may be far fewer desperate buyers.

Coercion is also always a concern. In general, people shouldn't be forced to use a new technology, perhaps especially where something as personal as reproduction is concerned. The good news on that front is that we seem to be a long way from the kind of thorough-going eugenics that would force people to edit their embryos or, presumably, not reproduce. We should not be overly confident about that. About half the American states forcibly sterilized about 60,000 people in the 20th century in the name of eugenics.[3] The Nazis gave eugenics such a bad name, along with its unscientific and coercive application in the United States, that it seems not now a threat. But it needs to be watched.

So does a more subtle form of coercion, an implicit kind of coercion. Sometimes the culture changes in ways that effectively force people to use a new technology, whether they want to or not. Urban (and suburban) development in most parts of 20th-century America made it very hard, in most places, for people who could afford it to avoid buying cars. I myself resisted first cell phones and then smartphones for several years until their value, particularly in family communication, effectively forced me to use them. (Well, their value and my wife.) We also, though, seem a very long way from germline genome editing becoming so common as to be implicitly required.

Human germline genome editing can certainly raise questions of equity. If the procedure is expensive, as it no doubt will be, at least at the beginning, it will not be evenly available. The rich often have access to things the poor do not, but often they are luxuries, not genetically healthy or enhanced children. Of course, the rich do usually have healthier children and children whose opportunities are often enhanced by what their parents can give them in completely nongenetic ways. For the next few decades, as the next chapter points out, there is not that much benefit (almost) anyone's children will be able to get from germline genome editing. The rich will not have superbabies, not because they won't want them but because no one knows how to make them—and won't, for a long time, if ever.

We should monitor the uses of the technique and how important or useful they actually are, as well as access to them. At some point, justice may well require us to guarantee access to the technology for at least some prospective parents. (I discuss this point a little more in chapter 15.) I cannot say that I am as confident that the requirements of justice will be met; safety is a much more politically successful goal. This concern is mitigated, though, by the fact that for the near future, the importance of access to human germline genome editing is trivial compared with the importance of access to, say, clean water, good food, and decent primary health care. Nonetheless, these possible inequities and their significance need to be watched.

Diversity is another important concern. Will human germline genome editing homogenize the human population, depriving us of people whose genomes lead them to make unique contributions to the rest of society—or, whether they help "us" or not, have ways of life that many find worthwhile. We already see some people in some groups who are widely considered "disabled" fighting back against attempts to "cure" them, let alone to "prevent" them, from the deaf, to people with achondroplasia and other little people, to some people with autism symptoms that they believe to be useful "neurodiversity." Germline genome editing, whether enforced by a government or just by parental decisions, threatens that kind of diversity.

I am not sure there is a good answer to those concerns. Some tension between people with unpopular conditions or traits and parents who want to avoid those conditions or traits in their children may be inevitable. We might consider increased support for people in those communities or perhaps improving their ability to make their case to prospective parents about the values of their conditions, and their lives. Or we could ban editing for those particular conditions. Would we consider banning all germline genome editing because of this tension? I suspect not.

The diversity argument also has a deeper aspect. If too many people edit their children to too similar genomes, the human species could lose diversity that might be crucial in future challenges. This too seems to me not something to lose sleep over soon. With 7.5 billion people, many with very different lives and preferences, even if germline genome editing became quite common, it will be a very long time before our genetic diversity is so dangerously reduced as to put the species at great risk. Nonetheless, it is possible that some genetic variant that was edited out will turn out to have been protective against some new environmental challenge (from global warming to a new strain of influenza). That risk too should be monitored, although the possibility that the editing could be reserved, either in the next generation or in the existing ones through somatic genome editing, mitigates this risk somewhat.

And finally we come to enhancement. This, I think, is at the heart of many of the greatest concerns about human germline genome editing. To some extent, that is a result of fears of economic or other inequities, with some people getting superbabies that others cannot have. But the concern runs deeper, to a worry about a change in the nature of humanity.

Not everyone worries about enhancement.[4] And, indeed, most of civilization is made of things that enhance us or our abilities. Binoculars enhance our vision, computers enhance our memory and communication, and airplanes enhance our mobility. One might argue that enhancement may be the most human of traits—or, at least, something we have been doing consistently since we started using fire and stone axes. Perhaps biological enhancements are different, though I am not sure how.

The primary question here, though, is not the hard question of whether biological enhancement is good or bad, but whether, if bad, it could (and would) be usefully regulated. Chapter 15 has some suggestions for such regulations, but enforcing them would be difficult, even apart from the politics of getting them adopted. Of all of the issues in this chapter, enhancement seems to me the one likely to be hardest to control. It is good that, as discussed in the next chapter, it is far from any meaningful reality in the next few decades even with human germline genome editing. But, in the long run, it may create the best case against the technology—at least, among those opposed to enhancement.

As we have run through these arguments, I hope you heard echoes of other arguments about technologies. All of these arguments (and a few more) have been applied to IVF and PGD; to computers and the Internet; to television, and, most likely, to reading and writing. They tend to be heard whenever novel technologies arise. That does *not* mean that they should be ignored; quite the contrary, they should be considered, debated, and monitored. But it should at least lead us to some recognition that we have managed to muddle through, not perfectly but not catastrophically, in the past. And perhaps some confidence that we could, if we choose, muddle through human germline genome editing.

13 Just How Useful Is Human Germline Genome Editing?

Not very useful in the near- to midterm; it's less clear in three or four decades.

I see no inherent or unmanageable ethical barriers to human germline genome editing. On the other hand, I see very few good uses for it. That is mainly because other technologies can attain almost all the important hoped-for benefits of human germline genome editing, often with lower risk. Two such technologies are particularly noteworthy: embryo selection and somatic genome editing.

This chapter looks at the surprising paucity of plausible useful functions for human germline genome editing, at least in the next several decades. I will specifically examine its potential for editing *out* disease-promoting genetic variations; editing *in* disease-preventing variations; and editing for enhancement, including ultimately to create new human species. As we look at each of those, remember that safety issues will remain for human germline genome editing for a decade or more, which not only pushes its possible uses off into the future but provides more time for alternative approaches to reach the same goals.

Editing Out Disease-Promoting Variations

The most obvious potential benefit would be to edit embryos, or the eggs and sperm used to make embryos, to avoid the births of children whose genetic variations would give them a certainty or high risk of a specific genetic disease. And here it is time to explain the ways genetic diseases or other traits get inherited. If the disease or trait depends on just one gene, we call it a Mendelian condition or trait, named after Gregor Mendel, the Austrian monk who first discovered this kind of inheritance. If more than one gene is involved, we cleverly call them non-Mendelian conditions or traits. Most of the discussion below is about Mendelian conditions for the simple reason that there is more to say about them.

Mendelian conditions can largely be put into five main categories, depending on where the relevant DNA is found and how many copies of the disease-causing variant are needed to lead to the disease: autosomal dominant, autosomal recessive, X-linked, Y-linked, or mitochondrial.[1] Autosomal dominant diseases require only one copy of the disease-causing genetic variation; autosomal recessive diseases require two copies, one from each parent. X-linked diseases typically require two copies in women (one from each parent) but only one in men (who have only one X chromosome, always inherited from the mother). Y-linked diseases, which are unusual, are found only in men and require only one copy—because only men have a Y chromosome and normally they have only one copy of it. Mitochondrial diseases are inherited only from the mother and any mother with the disease will necessarily pass it on to all her children.[2]

So, if an embryo has 47 CAG repeats in the relevant region of its *huntingtin* gene, it is doomed (if born) to have autosomal dominant Huntington's disease. One might use germline editing to reduce those 47 repeats to a safe number, of under 37, and thus prevent the disease. Or if an embryo has two copies of the genetic variation for the autosomal recessive Tay-Sachs disease, it could be edited so that the embryo had one or no copies and would be safe. The same is true of X-linked, Y-linked, or mitochondrial diseases.

If this is safe and effective, it may make sense. But another technology that has been in clinical practice for about 30 years is known to be (relatively) safe and effective and can do the same thing—PGD. PGD involves taking one or a few cells from an ex vivo embryo, testing the DNA in those cells, and using the results to determine whether or not to transfer that particular embryo to a woman's uterus for possible implantation, pregnancy, and birth. The first PGD baby was born in 1990.[3] In 2016, the last year for which data are available, the U.S. Centers for Disease Control and Prevention (CDC) reported that about 22 percent of the roughly 260,000 IVF cycles performed that year in the United States involved PGD (or a version called preimplantation genetic screening, or PGS).[4] That was up from about 5 percent the year before.[5] Anecdotally, from conversations with people working in IVF clinics, it sounds as though PGD or PGS usage in 2019 may well be above 50 percent, at least in some areas of the United States.

If a couple wants to avoid having a child with a nasty Mendelian genetic disease or condition, they could, in a decade or more, use CRISPR or other gene-editing tools to change an embryo's variants into a safer form or, *today*, they could use PGD to find out which embryos carry, or do not carry, the dangerous variants. For an autosomal recessive condition, on average 25 percent of the embryos will be affected; for an autosomal dominant one, 50 percent will be. Even for dominant conditions, if one looks at 10 embryos, the chance that all 10 will have the "bad" version is one in 1,024. If you have 20 embryos to examine, it becomes one in 1,048,576.

So, why take the new, riskier—and, to many people, disconcerting—path of gene editing rather than just selecting embryos? There are at least three plausible answers, though none applies to many people.

First, some people believe that embryos have a high moral status and should, if possible, be saved rather than discarded. From that perspective, a couple would make one or a few embryos, test them all, edit all the ones with genetic problems, and, over time, transfer all of them for possible implantation and birth. One problem is figuring out who, exactly, would put such a high a value on embryo protection while, at the same time, being willing to undertake the new, risky alternative of genome editing, with its aura of "playing God." Devout Catholic couples, for example, would disapprove of destroying or wasting embryos, but their church also teaches that any form of reproduction using something other than heterosexual marital sexual intercourse is sinful. Embryo editing would require IVF, in order to have the embryo available; PGD, in order to have the embryo tested to see if carries the dangerous variants; CRISPR, in order to edit the embryo's DNA; more PGD, in order to make sure the edits worked properly; and finally mechanical transfer of the embryo to the uterus. All of these steps are against Catholic doctrine. Perhaps some couples exist who value embryos so highly as to be unwilling to "reject" any of them but who, at the same time, are willing to go through the massive technical interventions of human germline genome editing. I suspect there won't be many.

Second, some couples may be limited in their ability to make enough IVF embryos for selection through PGD to be effective. If a couple makes only one or two embryos, the chances are higher that one or both will carry a risky gene. Of course, multiple rounds of IVF might provide further embryos, but some people will still not have a single "safe" embryo even among 10 or more. It's not only that the low odds sometimes happen (use IVF in a million women and 1,000 will have a once in a thousand outcome), but additional risk factors come into play; for instance, although they make, say, 10 embryos and five carry the "safe" genetic variations, all five of those may carry the wrong number of chromosomes and not be viable.

This may well affect some couples—at least for a few years— though we do not know how many. There is evidence that PGD, when used to avoid genetic disease, fails to produce live births in most cycles.[6] If the prospective parents try again, and possibly again and again, the failure rate should be lower (especially if it is done when the woman is relatively young, with a relatively larger production of eggs). In the longer term, we should be able to make eggs (and, if necessary, sperm) from skin cells, through an existing process called induced pluripotent stem cells. This has already been done successfully in mice, with both eggs and sperm. I have argued that it is likely to be in widespread clinical use in humans sometime between 2035 and 2055.[7] How soon will we have evidence for the safety of human germline genome editing that is good enough to allow its clinical use? If this timing holds, creating gametes through induced pluripotent stem cells may be available by the time, or not much later than, we have good evidence for the safety of human germline genome editing. That would eliminate a shortage of embryos as a problem with PGD.

This is not to say that the shortcomings of PGD may not justify human germline genome editing. That depends on several issues, among others the proven safety of human germline genome editing, the long-term success of PGD for couples, the proven ability to provide safe stem cell–derived eggs, and many others. It is possible that human germline genome editing might, for some period of time, be a proven better alternative for some people. My own guess is that this will not be the case, at least not for very long or for many people, but we will have to wait and see. It certainly will not be true soon.

The third and, as far as I am concerned, only truly compelling situation, is that of people who want to have genetic children but who cannot use PGD to select "safe" embryos because they do not have any safe genetic variations to pass down. Consider a couple where each has cystic fibrosis, an autosomal recessive disease. Any genetic child of theirs will get one bad copy from the father and one bad copy from the mother—because bad copies are all the parents have to give. Similarly, people with two copies of a genetic variation for an autosomal dominant disease can only produce genetic children who have at least one bad copy of the gene because the children *must* get a copy from them— they only have bad copies to give. These couples (recessive) or individuals (dominant) cannot make any healthy embryos or have any healthy genetic children without genome editing. The stories with X- and Y-linked diseases are similar, though more complicated.

Embryo editing might be able to help in these four cases, but also might not, at least in all of them. Some work by Shoukhrat Mitalipov's group[8] suggests that replacing a homozygous genetic variation (one where both copies are the same) may be difficult. Working with heterozygous variants, they concluded that the "replacement" copy for the "bad" variant came from the good copy on the other chromosome. If, as they infer, CRISPR gene replacement only works when the fertilized egg already contains one "good" copy of the gene, it could not be used to fix embryos that had two bad variants, such as those made by couples with the same autosomal recessive disease. Paul Knoepfler has put this forward as an argument against this use for human germline genome editing.[9] But the Mitalipov finding remains *extremely* controversial scientifically with lots of doubters. Even if it proves true, researchers may well find ways to circumvent it.

In addition, women with mitochondrial diseases also cannot use PGD to select healthy embryos, as all of their embryos will inherit their "bad" mitochondrial DNA. But for them, another solution, "swapping" mitochondria rather than editing its genome, is already being tested and may well be shown safe and effective long before human germline genome editing would be.

It is possible, of course, that human germline genome editing may be so wonderful that it will avoid the decades-long uncertainty about safety and efficacy that has characterized so many biomedical advances, from monoclonal antibodies to gene therapy. I cannot say it is impossible, but, personally, I wouldn't bet on it.

These examples require that the people with the disease live long enough and be healthy enough to be willing and able to have children. People born with many rare genetic diseases never live to puberty and will never find themselves in this position. Similarly, a (very) few people have two copies of a powerfully autosomal dominant disease variation. How many such couples are there in the world? That's impossible to know, but it seems highly unlikely it will be more than a few thousand out of the world's 7.5 billion people (and a large number of those will not have access to the high-tech treatment of CRISPRing embryos). Still, are speculative or metaphysical concerns about human germline genome editing good reasons to stop eager people, even if only a handful, from having a plausible way to produce healthy children from their own, otherwise unhealthy genomes—especially in societies that jump through hoops to make it possible for people to use IVF or PGD to have healthy genetic children? I think not.

But PGD is not the only technical possibility for avoiding disease in a child. In many cases of Mendelian diseases, it should soon be possible to "cure" the disease after birth (or possibly during fetal development) by editing the baby's (or fetus's) DNA—and often by editing only the tissues affected. Thus, rather than edit an embryo to avoid sickle cell, one could edit the erythropoietic stem cells in the bone marrow of a newborn or in a fetus to "cure" the disease. This kind of somatic cell gene therapy[10] has been under development for nearly 40 years. The FDA approved the first somatic cell gene therapy treatment for an inherited disease at the end of 2017,[11] but by 2025 it expects to approve 10 to 20 gene or cell therapies each year.[12]

Somatic cell therapy does not change the germline, and it comprises a technology much closer to being shown safe and effective than human germline genome editing. Arguably, the fact that the change is only being made in one or a few of the many tissues of the body would improve its safety over a change that exists in every cell, including cells where a particular off-target change has harmful effects.

On the other hand, genome editing of an egg, a sperm, or a zygote needs to change only one cell. This might prove more effective than changing, say, 100 million blood-forming stem cells or several billion lung cells. Furthermore, somatic cell editing would not necessarily work for all conditions. For some, too many different cells or tissues may have to be targeted. For others, the damage may begin before birth, or even before the stage of fetal development where in utero somatic editing becomes plausible. For diseases with very early consequential effects, somatic cell therapy may be inferior to embryo editing or embryo selection.

Even when somatic editing is possible, human germline genome editing retains one advantage: the process would not have to be repeated in the next generation. If somatic editing is used, that person would still have eggs or sperm that could pass on the disease. If she or he wanted to avoid a sick child, PGD or somatic cell gene therapy might be necessary. If germline editing is used, that child's children will be free from the risk of inheriting the disease from their edited parents. But is this a bug or a feature? It adds a choice—not a choice for the embryo that is, or isn't, edited but for the parents of that embryo. Somatic cell editing continues the possibility of a disease in the next generation—but allows that generation's parents to make the decision. One might—or might not—see that as a benefit.

Right now, one of the biggest drawbacks to somatic cell gene therapy is the extraordinarily high prices announced for the few approved therapies. Note that these are "prices," not "costs"—we do not know how much gene therapies cost companies to make and provide, let alone on what accounting basis (and accounting is the darkest of the dark arts). Perhaps human germline genome editing would be cheaper—a result that would depend on vagaries of patent protection and regulatory exclusivities that cannot now be calculated. If the human germline genome editing therapies are produced by the existing biotechnology and pharmaceutical industries, I think the possibility that they will be offered at lower prices than those for gene therapies, although a result devoutly to be wished, is highly unlikely. On the other hand, lower prices for gene therapies may be more likely, both as a result of a backlash against $2.1 million one-shot treatments and, perhaps, as a result of competition.

In non-Mendelian (sometimes called multigenic) diseases, no one variant plays a powerful role in causing the disease. Variations in two, or twenty, or two hundred genes may influence the condition. Collectively, those influences might be 100 percent, though the cases we know now add up to much lower certainties. We do not yet know of many good examples, though at least one paper claims to have found strong evidence that variations of different genes, working together, increase the risk for some cases of autism.[13] And, more generally, we know of many combinations of shared genomic regions that (slightly) increase or lower the risk for various diseases or traits in particular, studied populations. (These have led to the hot area of "polygenic risk scores," whose ultimate significance remains to be seen.[14])

The biggest problem with human germline genome editing for non-Mendelian conditions is that we do not know nearly enough about the conditions. We believe that many conditions are non-Mendelian, but how many genes are involved? Which genomic variations add or subtract risk? How do the effects of variations from different genes combine to create risks? In a simple world, they would be additive: if having a particular variation of one gene increases a person's risk of a disease by 10 percentage points and having a particular variation of a different gene increases that person's risk by 5 percentage points, then having both would increase the risk by 15 percent. But there is no inherent reason nature has to work that way; the combined effects may be greater or less than their sum. It is even conceivable that having two variations that each, individually, raise a person's risk might somehow lower the overall risk. We know almost nothing about the structure of these non-Mendelian, or multigenic, risks.

It is clear, though, that, in general, PGD would be much less useful for non-Mendelian diseases than for Mendelian ones. The chances of finding an embryo with "the right" set of genetic variations at five different spots along the genome will be much smaller than of finding an embryo with just one "right" variation. If the odds for any one variation are 50/50, the overall odds for any five variations in one embryo are one in 32. If gene editing could safely and effectively edit five places in an embryo's genome (or in two gametes' genomes), it could deliver the preferred outcome. On the other hand, if we can use genome editing to do that in an embryo or gamete, we may well be able to do the same in a fetus, a baby, a child, or an adult through somatic cell gene therapy—unless the condition begins to cause harm early in development, or broadly enough in the body that it needs to be delivered to all the body's cells.

Right now, there is no non-Mendelian condition for which we are confident we know the exact set of genes involved. Neither do we know the negative and positive effects of different combinations of genetic variants. Until these uncertainties are adequately resolved, human germline genome editing, though in theory better than PGD, will not be safe or effective enough for use. Once they are resolved, in many situations it will be no better than somatic cell genome editing, except for the possible absence of needing to hit targets in multiple tissues or cell types and the absence of a need to repeat the editing for the next generation.

Editing in Disease-Preventing Variations

Somewhere in the no man's land between treating disease and enhancing traits lies editing disease-prevention genetic variations into a person's genome. Instead of editing out an unusual pathogenic variation in favor of a common variant, the procedure could be used to edit out a common, normal-risk variant and turn it into an uncommon (or even rare) variation that lowers the risk below the population average. Two examples may be illuminating.

One is what He Jiankui tried to do: changing the normal sequence of *CCR5* to a variation with a 32-base-pair deletion that does not make a functioning protein. About 1 percent of Northern Europeans inherit two copies of this dysfunctional version[15]–its frequency is much lower in most of the world's population—which decreases the susceptibility of T cells to HIV infection.

The other example would change the normal version of a gene called *PCSK9* to a nonfunctional version. The normal version is involved in recycling low-density lipoprotein (LDL). In 2006 researchers discovered that a few people carried a version of the gene that made a nonfunctional protein. Those people had very low LDL levels and very low coronary artery disease risks.[16] (One very healthy 40-year-old had LDL levels of 14, compared with the recommended, but frequently unmet, goal of below 100.) Since 2015 FDA has approved two drugs to block the activity of *PCSK9* in people with very high LDL.[17] People associated with He's lab had inquired of various scientists about the activity of *PCSK9* and its different variants; he appears to have considered editing an embryo to turn the normal version into an inactive one.

PGD would almost never be helpful in choosing embryos with these protective variants. PGD can only pick embryos with variations their parents have. For these traits, inherited as recessive conditions because of their loss of function nature, each parent would need to have at least one copy of the variant. If you are not among the 1 percent of Northern European couples where each of you carried this specific *CCR5* variant, then no amount of PGD can produce an embryo that inherits this variant.

Genome editing should—eventually—have little to no trouble with these kinds of changes when they are Mendelian and involve only one gene. This is what He Jiankui tried—poorly, recklessly, and unsuccessfully—to do. But, eventually, once CRISPR or other gene editing technologies are proven out, this should be feasible.

Could somatic gene therapy work equally as well? As with many efforts to edit out disease-predisposing genetic variations, yes. Presumably, if this could work with embryos, it will work with living people. Again, embryo editing would work better than somatic editing if the problematic conditions started early enough so that even fetal cell therapy would be too late, or if the resulting change needs to be in a large percentage of the body's cells. On the other hand, as noted above, somatic cell therapy can try to target particular tissues, avoiding risks that might arise in other tissues, either from the planned change or from unplanned changes. Either type of change could have particularly damaging consequences during embryonic or fetal development; somatic therapy would minimize those.

The real problems with introducing uncommon disease-prevention variations to a person's genome are not with the process but with the result. Just what are the effects of changing a variant found in the vast majority of people (and, in some cases, of all mammals) to a version found only in a few? With both *CCR5* and *PCSK9* null variants, we know that some people homozygous for them end up as healthy adults. But how many? The fact some people make it to healthy middle or old age says that the functional gene is not essential for a healthy life, but it doesn't tell the whole story.

So far, no one has, for example, done a long-term study of a large group of babies who have these variations, and who come from various ethnicities or from societies with different dietary and environmental conditions, to see what happens to them. As noted in chapter 9, in June 2019, though, one group did publish an article based on looking at the *CCR5* status of more than 400,000 people in the U.K. Biobank and finding that those with two nonfunctional, and thus somewhat protective against HIV, variants died sooner, but the authors retracted that paper after a technical error had been pointed out to them.[18] There are, however, other studies providing evidence that people without a functional *CCR5* gene have much worse symptoms and outcomes when infected with influenza[19] and West Nile Virus.[20] Before changing people over from common to rare variants, we need to be confident we know as much as possible of the negative and positive effects of those rare variants, as well as the positive effects of the common ones.

And there is one other qualification to using human germline genome editing to give a person a much lower than average risk of disease. I, along with almost all living humans, have a very, very low risk of contracting polio—not because of genetic changes (except in some of my immune system's B cells) but from a vaccine. My chance of dying of cholera is very low not because of genetic changes but because of public health measures. Even if I were to contract cholera (at least if I had access to good medical care), I would benefit from many treatment options given the advancement of medical knowledge over the past century. Even my favorite example of a couple with an autosomal recessive disease, cystic fibrosis, has in recent years seen the development of very exciting, and potentially greatly life-extending, treatments that are pills, not gene therapies.[21] Why would we use germline gene therapy to introduce resistance to polio or cholera, instead of other preventive or treatment measures? It all comes down to balancing risks and benefits. If gene editing as a preventive measure is much better than anything else *and* it is known to be safe (and inexpensive), it may make sense. Eventually. But just because we *can* use gene editing to prevent a disease doesn't mean we *must* if there is a safer, more proven, or cheaper technology already at our disposal.

Editing for Enhancement

What about editing for frank enhancement—not for giving people "enhanced" protection against diseases but for giving them better than normal traits or abilities for sports, math, music, beauty, school, or anything else. Enhancement is at the heart of the popular fears of human germline genome editing. This section discusses some of the scientific merits and demerits of enhancement through human germline genome editing, building in part on ideas I have expressed before about human biological enhancement in general.[22] The question whether such enhancement, if safe and effective, would be ethical was discussed in chapter 12.

The biggest technical problem with enhancement through human germline genome editing is that we know of almost no genetic variations that clearly are enhancing. It seems fairly clear at this point that there are (almost) no Mendelian enhancement genes,[23] even taking a wide view of enhancement to include things like hair color. Some single genes make significant contributions to things like hair color, eye color, and skin color, but along the lines of dark versus light, not light blue versus deep brown. (One of the few exceptions is a gene that, when mutated in a particular way in both copies in a person, generates a particularly strong form of red hair, complete with freckles—and also a much higher risk of the dangerous skin cancer, melanoma.[24]) Although cosmetic traits have not been heavily explored, and genetic associations have, no doubt, yet to be explored for all possible kinds of "enhancement" (Scrabble ability?), it seems highly unlikely at this point that many powerful Mendelian genes will be found.

This means PGD is all but useless for enhancement, and it means both human germline genome editing and somatic cell editing for enhancement would be quite complicated, though perhaps not impossible. (It may also make some nongenetic methods of enhancement relatively more attractive, like hair dye, colored contact lenses, or intensive Scrabble training.) Knowledge of some sets of genomic variations that enhance traits may ultimately come from the simple availability of more genotypic and associated phenotypic data. If we had the genomes of the world's 1,000 top Scrabble players, maybe some genomic associations would be found. Knowledge of other causal factors may, or may not, come from research into diseases or conditions that cause well below average performance, which may provide some clues about DNA locations that might be associated with better performance. At this point, though, with the possible exception of the uncommon disease-preventing variants discussed in the previous section, we have almost nothing to offer that is arguably enhancing.

Even if we do, we still will need to be careful about the safety and efficacy of that combination of variants. If, for example, having one of two equally common variants of 20 different genes produces extraordinary performance, that makes those combinations roughly one in a million (assuming mating that is random to those traits and various other provisos). We may know some people with that combination of variants who are tremendously successful in a particular way—but do we know the life, and disease, courses of enough of them to know that it is safe? As with the uncommon variants discussed in the previous section, uncommon combinations of genomic variations also raise questions of safety. They may also raise questions of efficacy. A spectacular Scrabble genome will do no good depending on the rest of the genetic and, importantly, cultural context. A combination that works in one ethnicity may not work as well, or at all, in another one. And people who never learn to read; who only read character-based, not alphabet-based, languages; who never encounter Scrabble; or who never have time to play it will *never* be Scrabble champions, regardless of their genomes.

If we develop knowledge of safe and effective combinations of variants that produced prized characteristics, we still have to ask whether human germline genome editing is any better, or worse, than somatic cell editing, or than nongenetic methods of producing the same result. Endurance athletes can have high hemoglobin levels from transfusing themselves with red blood cells before a race, using erythropoietin (a drug that increases red blood cell counts), living and training at high altitude, or sleeping in a low-oxygen tent.[25] They could (probably) get higher hemoglobin levels by doing somatic gene editing to increase the number of erythropoietin genes in the relevant cells, or how regularly they are turned on. One might do the same thing by human germline genome editing, giving the future athlete this "enhancement" at least from birth and possibly before. Or they could "win the genetic lottery" and be born with genetic variations that naturally produce high levels of hemoglobin, like the exceptional Finnish cross-country skiing champion, the late Eero Mäntyranta,[26] or less rare Tibetans, with genetic variations adapted to low oxygen levels.[27] (We should also note that higher levels of hemoglobin and red blood cells are also associated with strokes and other diseases. Although we don't have definitive evidence of causation, more than 35 bicycle racers have died, in three clusters, in ways that seem possibly related to hemoglobin "enhancement."[28])

The International Olympic Committee bans transfusions or erythropoietin use. And it is quite worried about somatic or germline genome editing and wants to ban people who have used them from competing. All the others are perfectly legal under its rules.

And, finally, at the far end of "enhancement" lies the science fiction fear of a speciation event—humans separating into two (or more) different species through someone's creation of "*Homo superior*," "*Homo sapiens 2.0*," "*Homo deus*,"[29] or "the Eloi" and "the Morlocks."[30] Does human germline genome editing make this possible?

Note first that there is a difficult definitional question here— just what is a species? Philosophers of science continue to juggle many different and somewhat contradictory definitions.[31] In many concrete cases, controversy continues to rage about whether two populations of living organisms are the same species, the same species but separate subspecies or "breeds," or are two different species. Some of these taxonomical disputes concern other humans—or, at least, other entities generally accepted to be part of the genus *Homo*. Whether to call Neanderthals a subspecies of *Homo sapiens* or a separate *Homo neanderthalensis* has been controversial. Similar issues may plague more recent humanlike species, such as the Denisovans or the hobbits (*Homo floresiensis*). And clearly drawing readily accepted species lines between various earlier members of the robust bush of human relatives has proven extremely difficult.

One often used measure of species is the ability to interbreed—or, perhaps, in light of zoo-bred tigons and ligers, a propensity to breed together in the wild. (Even that reservation doesn't save the definition, as wolves, coyotes, and dogs—all classed as different species of canids—interbreed in the wild, as do polar bears and grizzly bears.) If accepted, though, this would have implications for how we understand the human species, as the majority of living members of *Homo sapiens* carry genetic sequences derived from Neanderthals and Denisovans as a result of interbreeding between those human groups and our ancestors. Would that mean that all three groups have to be considered one species?

Still, if one wants to use that definition, it might well be possible for a clever scientist to engineer human embryos genetically so that they could only have children (without resorting again to gene editing) with similarly modified humans, perhaps by changing some of the characteristics of eggs and sperm or, as Lee Silver suggested over 20 years ago, by changing in the number of their chromosomes.[32] Such people could be, in every respect except their potential reproductive partners, exactly the same as the rest of humanity, but, by the interbreeding definition, this would be a speciation event. That said, why would anyone want to do it?

One might instead mean by a speciation event a change that produced living people who were substantially different from contemporary *Homo sapiens* in phenotype or in genotype—as different, or more, as Neanderthals were from us. That is also plausible, sometime, but no time soon. Just as we know next to nothing about genetic variations that cause more limited enhancement, we know even less about variants with such great effect as to make an entirely new species. One might try randomly throwing into human embryos variants partially or even entirely from other animals (or plants or bacteria), or freshly designed "new and improved" genomic variations or genes, and hoping that some of them would survive until birth. Again, though, why would anyone want to do that without having a good idea of what kind of being would result?[33]

And, of course, at the end of the day we are still left with the question of whether having more than one human-derived species would be wrong. It could become a bad thing if it led to discrimination, warfare, or genocide, but would a peaceful pluralism of species of the genus *Homo* be a bad thing? Or, for that matter, the eventual replacement of *Homo sapiens* by one or more successor species? There is no reason to think our species will exist forever. And while we should be concerned about the individuals in our species who are alive, is there any reason other than a sentimental attachment, akin to a rooting choice among sports teams, to prefer our species in some indefinitely distant future? (I, for one, would not try to base such a preference on the "goodness" of *Homo sapiens*, at least not based on our historical record.)

In any event, these questions need not trouble us today. If (or when) they will arise, it will be the future, probably the far future, and we will all be dead. Our descendants will make up their own minds, no doubt little caring for our ancient thoughts.

The point to remember from this chapter is that human germline genome editing, even if it turns out, eventually, to be safe and effective, has very few plausible good uses. That's true whether we look at editing out disease risks, editing in disease prophylaxis, doing frank biological enhancement, or trying to split our species. For these uses, we know little about its efficacy or about its risks. Often we do not know enough about genetics to know how to use it even if we thought it was proven, in general, safe and effective. And for those things it might be used for—the ones we largely agree on, such as disease avoidance and prophylaxis, and the ones that are much more controversial, such as enhancement or speciation—there are current alternatives, especially PGD and human somatic genome editing, that are more available, better understood, and in most cases likely to be just as effective.

14   How to Test Human Germline Genome Editing

Very carefully.

After reviewing the He Jiankui experiment and then the broader implications of human germline genome editing, what *should* be done? First, and foremost, *not* He Jiankui's experiment. That effort was wrong on an impressive number of levels. But what would a good process look like that might lead to good decisions on human genome germline editing? Ultimately, the answer is whatever process or processes, if any, a particular jurisdiction chooses; what "we" want depends importantly on who the "we" is.

But I do have some suggestions. The process will need to answer at least three questions: Do we want to allow any human germline genome editing, do we want to allow human germline genome editing for only certain uses, and how would we test it for safety? And, of course, there is the "metaquestion" of the process we should use to answer those three questions. Part of me wants to take the questions in that order, but, perhaps counterintuitively, it seems to me the questions of "any" human germline genome editing and, if so, to what ends, might be better approached after a consideration of how and how well, if at all, any safety can be ensured. This chapter will try to answer the safety question; the next chapter will handle the others.

Human germline genome editing must only be adopted if there is strong and rigorous evidence that it is reasonably likely to be safe for any babies who result. Note that I do not ask that it be proven "safe." I don't think anything can be proven "safe," or anywhere close to safe, at least until it has been widely used for many years (and probably not even then). Good evidence about the safety of the process can and should be obtained before the process can be used in a clinical trial or approved for clinical use, with safety monitoring continuing after approval. But that is not enough. In addition to establishing the safety of an approach in general, one must work to ensure that safe approaches are implemented.

We will start with the ways to establish the overall levels of safety of human germline genome editing, ways that fall into two categories—preclinical trials, with human tissues and with nonhuman animals, and then clinical trials, with a small number of carefully selected human volunteers. We'll end by discussing a framework for establishing systems for safe implementation in clinical use.

Preclinical Trials

How can we know whether human germline genome editing is safe enough even to try in humans? Two different paths can be followed: studies of ex vivo human and nonhuman embryos and studies of in vivo nonhuman reproduction.

The ex vivo embryo studies would focus on how embryos react to germline editing. This is all research in a dish; none of the embryos will be implanted or even transferred into a uterus.

The first step is to see how well the editing process works in terms of changing the DNA sequence. If the editing happens at the embryo stage, edited embryos should be sampled at various stages, from immediately after the editing has been tried up (which may be as early as the time of fertilization) until the blastocyst stage, at five to seven days after fertilization.

One might want to sample once a day or perhaps after each (rough) cell doubling, maybe eight times. At every stage, each cell could be checked to see what changes the editing has wrought. How many of the cells got the appropriate edit? How many got an unplanned variation on the desired edit? Are there any off-target changes, and, if so, how many? Ideally, one would like to do whole genome sequencing on each of those cells.

Of course, in addition to the edited embryos, one would want to have a control group of unedited embryos, and perhaps of unedited embryos that had undergone some of the process involved in editing, whether that means injection, viral infection, or something else. And those control embryos should also be sequenced at regular intervals.

But more than the DNA sequence can and should be inspected. Different genes are turned on and turned off at different times in the development and life of an organism, whether the organism is an adult or an ex vivo embryo. This process can be observed through a method called RNA-seq, which looks at messenger RNA to see what stretches of DNA are turned on in ways that could lead them to become proteins or active small RNA molecules. This approach should also be applied with the edited embryos at specific times through the blastocyst stage and also with an adequate number of control embryos. That can reveal what DNA is being expressed during those early embryonic stages, not during the organism's later life, but even knowing whether gene expression for those first few days is similar to, or different from, that of nonedited embryos would be valuable information. The more that it is different, the greater the safety worries.

Mosaicism will be an important issue. Mosaicism, in a biological sense, is when different cells in the same organism have different DNA.[1] A great concern about embryo editing is that not all the cells of the embryo will have received the edit, or exactly the same form of the edit. So the point to analyzing these series of embryos at different points in development is not just to see what DNA sequence or RNA expression pattern they have, and how it compared with the unedited embryos, but how many different DNA sequences or RNA expression patterns they have, again compared with the unedited cells.

One hard-to-answer question is how many times a series of embryo edits should be repeated. Should the progression from a zygote to a blastocyst be analyzed once, ten times, a hundred times, or a thousand times? Surely, one or ten would seem inadequate to give a sense of the degree of variability. One hundred might be enough; one thousand might be too few—that may well depend on just how good the results are. But if each series involved 500 cells and the cost of doing both the whole genome sequencing *and* the RNA-seq was, say, $600 per cell (about the cost of current single-cell whole genome sequencing), doing the roughly 500 cells in one series would cost about $300,000, doing 100 series of 500 cells each would be $30 million, and doing a thousand, $300 million. That's a lot of money to spend for just one part of preclinical trials, but the price of single-cell sequencing will undoubtedly fall substantially . . . and, if the technology is truly important, even those costs may be worth bearing.

Now consider three more issues about safety testing on ex vivo human embryos: the availability of human embryos or eggs, the use of cloned embryos, and the use of postblastocyst embryos.

Testing lots of human embryos will require . . . lots of human embryos. Where will they come from? Currently much human embryo research relies on embryos created for reproductive purposes but then donated by the prospective parents for research use. Ethical guidelines, and in some places laws,[2] forbid compensating the prospective parents for their embryos. It is possible that enough embryos could be obtained from altruistic donations by prospective IVF parents, but researchers may not want to use them. Prospective IVF parents are not a cross-section of any place's adult population—they will typically be people who either have fertility problems or carry genetic risks. (On the other hand, those kinds of prospective parents may be the most likely to try human germline genome editing, in which case their embryos may be an especially useful sample.) At the same time, the donated embryos will not be a random cross-section of all of the parents' embryos. They will have used their "best" embryos to try to make their own babies; these will be the leftovers. And the processes used for IVF will vary from clinic to clinic in ways that may make researchers concerned about their ability to make good comparisons between embryos generated in different places.

A solution to this is for the researchers to make, under carefully controlled circumstances, the embryos themselves. In many jurisdictions that is, in itself, illegal;[3] in others, including the United States, funding for this is limited. (Under the so-called Dickey-Wicker Amendment, federal funds cannot be used in research that destroys or puts at risk human embryos.[4]) But even where it is legal, it has a problem. Making lots of human embryos requires lots of human gametes—sperm and eggs.

Sperm is cheap, easy, and safe to obtain in vast quantities. Egg harvest is expensive, unpleasant, and risky. To make hundreds or thousands of embryos will require hundreds or thousands of eggs. Where will they come from?

Currently, in jurisdictions that allow the creation of human embryos for research, those eggs usually are donated by the women from whom they were harvested, either as part of an IVF cycle (for themselves or as potential donor eggs for another woman) or purely for research. This also raises problems, though—some jurisdictions do not allow women to be paid for their eggs, even through a discount on their IVF costs,[5] and egg harvest is unpleasant enough that purely altruistic donors may be hard to find. Even California, a state very friendly to medical research, banned such payments until a bill passed in October 2019—and that bill imposes specific consent and approval requirements on such paid donations as well as limiting them to compensation for the donor's time, discomfort, and inconvenience.[6] (The recent California law repealed a prohibition on paying women for eggs for research—but not for egg "donations" for reproduction—that had passed in 2006 as a result of pro-life groups and feminist groups concerned about the possible exploitation of poor women.[7])

But rather than soliciting the egg donations, with or without compensation, as mentioned in passing in chapter 13, researchers might be able to make eggs. The process of making gametes from stem cells is sometimes referred to as "in vitro gametogenesis" (IVG). This method, using either embryonic stem cells or induced pluripotent stem cells, has already been done successfully in mice, and work is proceeding on making this available for people.[8] My 2016 book, *The End of Sex and the Future of Human Reproduction*, is based in large part on the assumption that this will become safe, effective, and widely available for humans in the next 20 to 40 years (or earlier).[9] The use of IVG to provide eggs (plenty of cheap sperm will, no doubt, be available through more traditional methods) could answer the problem of a possible egg (and hence embryo) shortage, but it would raise the question whether embryos made from these IVG eggs were identical (or close enough to identical) to naturally occurring embryos to be good comparisons. If they were used for this safety testing, research on their comparability would be required (although such research would probably have to be done in any event before such gametes were approved for clinical use).

A second issue might arise as an attempt to overcome the natural genetic variations in the embryos being tested. Even if all the eggs and sperm used to make the embryos that are tested came from the same parents, each of those embryos will carry a different combination of the parents' DNA. It might be that any differences in the efficiency and safety of the editing have to do with those inherited DNA differences. That could be countered by using cloned embryos. If one human embryo were cloned, say, 100 times, it could provide 100 genetically identical embryos for testing. (On the other hand, although this might be good for some purposes, that avoided genetic diversity could be very important in clinical uses.) Although a method for cloning human embryos was discovered in 2013[10] and quickly replicated, researchers have not rushed to study that method, its efficacy, or costs. And the process of cloning itself might introduce new and different risks or complications into the experiments, which might, or might not, be justified by the importance of testing genetically identical embryos. At this point, it is probably best to keep this open as a possibility, pending more evidence on both the importance of inherited variations and the progress of human embryo cloning.

The possibility of extended study of ex vivo embryos can produce another interesting variation. In the more than 40 years of research with ex vivo human embryos, those embryos have always died after six or seven days unless the blastocysts were transferred into a human uterus, where they successfully implanted. In 2016, two laboratories reported that they have been able to keep human embryos alive outside the uterus, without implantation, for past what had previously been an unbreakable nine-day barrier (and an almost always unbreakable six- or seven-day barrier).[11] Both have taken those embryos up to the edge of two weeks, tethered not to a woman's endometrium but to a plastic apparatus in a petri dish.[12] This has raised an ethical discussion; several ethics guidelines, starting in 1979, and some national laws, have said that research cannot be carried on in human embryos more than 14 days after the fertilization that led to them.[13] The timing was originally driven by the average age of an embryo when it forms the so-called primitive streak, the first part of the growing embryo that looks different, and that later leads to a different set of cell types. (Among other things, it is thought that the primitive streak also marks the end of when one embryo can divide and become identical twins, as such a division would likely leave one twin with the wrong parts of the now differentiating embryo.) The laboratories involved in the work followed the guidance and destroyed the embryos before they crossed the 14-day barrier.

The ethical discussion continues unabated, but this may raise an interesting issue of safety testing. Should the ex vivo safety testing for edited human embryos extend, through those methods, beyond the seventh day and up to (or beyond) the fourteenth? On its face, that seems to make sense—why not get even further information on whether the edited embryos develop normally? The problem is that no one knows whether those ex vivo embryos kept alive past the normal time for implantation are developing the same way normal embryos are. And it is not clear that evidence on that will become available anytime soon. After all, it is not easy to do, for example, RNA-seq on an embryo that has been implanted in a woman's uterus for less than a week. In theory, such further testing would make sense but only if we have a "normal" baseline against which to measure it.

So, whatever variations are used, let us say that we now have robust experimental results comparing the DNA sequence and the RNA expression of edited embryos to unedited embryos, probably having started with nonhuman embryos and then moved to human ones. How do we evaluate those results? If the edited embryos look the same as the unedited ones, presumably that would give us some evidence that the process may be safe. If the results are grossly different, that should act as a warning siren. But what if they are "slightly" different? How different is enough to decide not to proceed with clinical trials, trying to use edited embryos to make babies? That will be a difficult decision. Happily, it need not be made based solely on studies of ex vivo human embryos; studies of the safety and efficacy of using edited embryos in reproduction in other animals can shed some light on this question.

The ex vivo human pathway stalls, either at seven days or, possibly, a few weeks later. It cannot tell us even whether, how often, or how safely such an edited embryo will implant, let alone develop. And I do *not* think we should, or could, make women pregnant with embryos with the full intention to terminate those pregnancies at specific dates so that the embryos or, after eight weeks, fetuses, can be studied for the normality of their development. I do not have strong qualms about abortion, at least before viability, and have no qualms about research on pre-14-day ex vivo embryos, but the cold-blooded creation of apparently viable pregnancies with the express *intention* of terminating them for research appalls me—and, I am confident, would appall almost all Americans. Unless and until American culture, and that of many other countries, changes radically, these experiments will not happen.

I would note though, that this view may not be shared universally across the world. In some cultures, perhaps especially those where abortion has long been widely available and used as a form of "belated contraception," such experiments may not appear horrific. What kind of reaction the rest of the world should, could, or would take to such experiments is an interesting question, but one I will not discuss unless and until they become a real possibility.

Happily, alternatives, although imperfect, exist. We can study the progress of genome edited embryos in vivo through their prenatal development in nonhuman animals. Of course, not all animals will do. We would need to use animals with reproductive systems somewhat similar to ours—no egg layers, but placental mammals. They should be accessible, common, and not exorbitantly expensive mammals, easily handled in experiments . . . so no endangered species and no lions, or tigers, or bears. And they would have to be mammals for whom we can acquire enough information about their reproduction to make them good model organisms for reproductive experiments.

Mice and rats stand out. We have used them in laboratories for decades and understand a great deal about their reproductive processes. Other similar small mammals, such as hamsters, guinea pigs, or rabbits, might also serve. Some or all of such species should be used. Embryos from them should be edited and transferred into the uteruses of females ready to carry them. When pregnancies are established, they should be carefully monitored, including with invasive monitoring that might be too risky in human pregnancies. And, at predetermined times, the pregnancies should be terminated and remains studied. Again, this should be done both with genome edited embryos and with control embryos, to see whether and how the edited embryos really are different. Assuming the pregnancies proceed successfully enough, some should be allowed to go to term and the resulting animals—both gene-edited and control animals—studied throughout their lifetimes. (Another advantage to using small mammals is that their life spans are relatively short; a two-year-old laboratory mouse is ancient.)

But those species also come with one gigantic disadvantage. They are not just not human, but they are not particularly close to humans, in evolutionary time (the paths that led to humans and to rodents diverged about 100 million years ago) or in physiology.[14] An old and, to drug developers, increasingly *unfunny* joke says that we can cure all human disease—in mice. Rodents have proven to be good models for some human diseases or treatments, but they are terrible models for others, notably Alzheimer's disease.

Small mammal tests must be done. If they show problems, those results should weigh heavily against starting human trials. But if they show success, they cannot be taken as establishing safety. In some cases, they may be close enough to allow human trials to start, especially if the first research participants will be competent adults, the human disease is serious with very limited treatments, the treatment affects physiology we understand fairly well, and we have good reason to think the relevant systems are "fairly" similar between humans and rodents. None of that is true with human germline genome editing.

So, now what? Try species more similar to humans. And that almost certainly means nonhuman primates. This sounds good, but it is . . . complicated.

For one thing, although the term "nonhuman primates" can roll trippingly off the tongue, it is important to realize it covers a wide range of quite different species. Depending on who is counting, there are somewhere between about 200 and 500 living primate species, ranging in size from Madame Berthe's mouse lemur (no, I am *not* making that up), which weighs about one ounce, to the Eastern gorilla, which can commonly weigh up to 440 pounds. (This should remind us of the dangers of generalizations about a species—there are humans who weigh much more than 440 pounds, but they are not common.)

Many nonhuman primates can be quickly dismissed, leaving monkeys and apes. New-world monkeys (known, technically, in Latin as "flat nosed monkeys") split from the old-world primates about 40 million years ago. Some new-world monkeys are used to a significant extent as laboratory animals, notably marmosets, squirrel monkeys, and owl monkeys.

Most nonhuman primates used in research are old-world monkeys (in Latin, the "down noses" because their nostrils point down). About 25 to 30 million years ago, the ancestors of this group diverged from the apes (including humans—note how your nostrils point). About 130 to 140 species exist, making them the largest primate group, and they range in size from 2 or 3 pounds to 110 pounds. The most common nonhuman primate laboratory animals, accounting for about two-thirds of all nonhuman primate research at least in the United States, come from the macaque genus.[15] Rhesus macaques are dominant, but cynomolgus macaques (also known as crab-eating macaques) are also common.

That leaves our closest relatives, the other apes. The lesser apes comprise about 18 species of gibbons, rarely used as research animals. The great apes include eight living species in four genera: orangutans (three species), gorillas (two species), chimpanzees (two species), and humans (only one remaining species). It is important here to note the difference between invasive (usually biomedical) research and behavioral research, which continues, in the wild, in zoos, in sanctuaries, and (in some cases) in laboratories with most of the great apes. But of the seven nonhuman great ape species, only chimpanzees have been at all widely used in biomedical research. At one point about 1,500 chimpanzees were being used for medical research in the United States; the number had been ramped up through breeding programs for research into HIV until a better, monkey model was developed. Since the 2011 publication of a U.S. Institute of Medicine report on research with chimpanzees, their use in medical research has dwindled and N.I.H. support for such use has disappeared.[16] It is not clear that any chimpanzees are still being used for biomedical research.

If one wanted to try human germline genome editing on the nonhuman primate closest to humans, one would pick chimpanzees, either bonobos or the "common" chimpanzee. But that poses a problem. The closer an entity is to human, the more the ethical concerns that prevent us from using humans for particular research projects apply to that entity. (I have already pointed out this dilemma several times in other work in the context of neuroscience and various surrogates for human brains.[17])

Chimps are disconcertingly close to humans, so much so that the ethical, and popular, opposition to invasive research with them has nearly ended such research, as well as sparked efforts, successful in places, to have such research banned—and efforts, thus far unsuccessful, to confer at least some degree of human personhood status on them.[18]

But there are additional problems. Chimpanzees are big, with an average size of about 90 pounds in females and 110 pounds in males, but, more importantly, they are impressively strong. Though smaller than humans, chimps are about one-third to one-half stronger. A chimp could tear a human apart, which makes invasive research difficult. Presumably, egg harvest and then subsequent embryo transfer would have to be done under general anesthetic, as would any other invasive procedures, such as pregnancy terminations, blood draws, or perhaps even ultrasounds. Although a 1983 article heralds the first "successful" efforts at IVF in chimpanzees (efforts that did *not* lead to successful pregnancies), it seems, in opposition to the recommendations of that article, to have been rarely if ever attempted or, in chimps or any of the other nonhuman apes, to have succeeded.[19] The other great apes—orangutans and gorillas—are bigger, stronger, less well understood, and without any history of laboratory animal use. For a nonhuman great ape model, the answer seems to be chimps or nothing . . . which makes the likely answer nothing.

So, let's look again at the old-world monkeys. They are smaller than humans, and they are 25 to 30 million years "different" from humans, but we have successfully done IVF with some of them, including rhesus macaques, starting in 1984.[20] This first "test tube macaque" was still alive and healthy 15 years later.[21] And, even better, germline genome edited cynomolgus and rhesus monkeys have been successfully produced since 2014, notably in China.[22] And large colonies of both monkey species exist for research purposes. Although it is tempting, for something as potentially risky as human germline genome editing, to call for preclinical studies in the nonhuman animals most closely related to us—chimpanzees—the practical and ethical issues incline me to favor rhesus or crab-eating macaques.

Medical experiments on most animals, along with most research on animals, has its own regulatory framework, committees, known generically in the United States as Institutional Animal Care and Use Committees (or IACUCs). These committees are ordered to assess whether the specific interventions, laid out in detailed animal research protocols, may go forward. To do so, they compare the risks, including pain and death, to the animal subjects with the potential medical or scientific benefits. And they demand that those risks, and particularly any pain, be minimized as far as possible. IACUCs do not have jurisdiction over all animals but do rule any research using vertebrates, which comprise fish, amphibians, reptiles, birds, and mammals, including rodents and primates. (Not that anyone is going to think crucial animal testing of human germline genome editing could be done in worms, insects, or octopuses.)

Both the small mammal (probably rodent) trials and the nonhuman primate trials can only proceed with animal protection committee approval, IACUCs in the United States and their equivalents elsewhere. The nonhuman primate trials should only start if there are promising results on nonprimate, small mammal trials. If so, they should do the ex vivo protocols described for human embryos. If successful, those should be followed with the same approaches as in the small mammal trials, with intensive (and possibly intrusive) monitoring, regular terminations to take stock of development, and, if it works, careful and long-term follow-up on any infants born.

Clinical Trials

If, and only if, there are promising results from the human ex vivo embryo studies and the nonhuman in vivo studies (particularly in nonhuman primates), it may be appropriate to go forward with clinical trials—testing germline genome editing in humans with the goal of making edited babies. This is more than a technical decision; it requires a decision whether that jurisdiction wants to go forward with human germline genome editing as an option at all. I discuss this question in the next chapter, but a negative decision would prevent any clinical trials (and make pointless any nonclinical research).

Assume, for now, that a society has decided, either before or after the preclinical work, that it is willing to consider allowing sufficiently safe human germline genome editing. Given that decision, it will still need careful scientific and medical consideration of the risks and the potential benefits of any clinical research, research aims to create real human babies, and to put them at risk. Even in light of preclinical evidence, how does one weigh the risks versus the benefits? Is 99 percent "success" in rodents and 98 percent success in nonhuman primates, defined as healthy offspring, good enough? And does it matter for what purposes parents are seeking to use human germline genome editing?

Fundamentally, this is not a different question than that which the FDA or similar agencies must answer any time they decide to allow a first trial in humans. As discussed in chapter 5, FDA must issue an IND to allow research use of an unapproved drug or biological product. That can be nerve-wracking: one *cannot* know whether a treatment will be safe and effective in humans until it has been tried in humans; we have too many examples, old and new, of treatments that looked fine in preclinical trials but either did not work, or worked disastrously, when tried in humans. Regulators must look at the data, hard, and, if the decision is to go forward, hold their breaths until the results come in. And as also discussed in chapter 5, the decision to move to clinical trials has another regulatory check—human subjects research committees (IRBs in the United States). Thus, approval for clinical trials, at least in the United States, must come from the FDA, *and* from a human subjects committee, as well as from human subjects by their decision to participate.

But what would a clinical trial look like? Clinical trials are generally classed as phase 1, 2, or 3. A phase 1 trial is done with a small number of people (often under 30) and is an effort to see if the intervention is safe, with no, or little, regard to its efficacy. A phase 2 trial is larger, often including one or two hundred participants, and looks at both safety and efficacy. A phase 3 trial can encompass several thousand participants, as well as varying doses and other variations of the procedure. But these are generalities. Trials for rare diseases may have only a handful of people and may combine two or all three of the phases into one trial with 10 or 20 people. Trials for common conditions, such as high cholesterol, may involve 3,000 or more participants.

For human germline genome editing, with its unknown but potentially high risks to babies, I would start with a small number of subjects, probably about 20 couples. These couples would have to have powerful reasons to want (need?) human germline genome editing. Whether human subjects research is ethical depends on the balance between the risks to the participants, which in this case must include most importantly the potential baby, versus the benefits, to the participants and to medicine, science, or society more broadly. The potential benefit of, say, partial immunity to HIV infection is much less powerful in the balance than the benefit, to the potential child, of being born without a bad genetic disease (for the child of two parents who both have such a disease) and, to the potential parents, of never having their own genetic child. But the potential benefit of partial immunity to HIV infection is better than the benefit of a possible but uncertain few additional points on an intelligence test.

Something like 20 couples should be found whose genetic children would necessarily have genetic diseases unless their embryos were gene edited. The pregnancies should be monitored as closely as possible without putting the woman or the pregnancy at risk. And any babies born should have both their DNA and their general health and behavior monitored very closely, potentially for their entire lives. This would be, in effect, a combined phase 1 (safety) and phase 2 (safety and efficacy) trial. Its results should then lead to a decision about whether the approach should be abandoned, whether larger trials should be performed before a final decision is made, or whether the first trial was enough to approve the procedure.

Approval and Afterward

Unless the early results lead to a decision to abandon the effort, at some point someone (in the United States, the FDA) will have to decide whether human germline genome editing is safe and effective enough to be allowed into general, nonresearch, clinical use. Like the decision to go forward with the first-in-human clinical trials, this decision cannot be made by an easy "fill in the blanks" algorithm, but will, except in extreme cases, require tough weighing of uncertain risks and benefits that are not commensurable—that cannot be reduced to the same terms. How much benefit, with how much confidence, is worth how much risk—and what is the role of the government in telling presumably competent adult prospective parents how that risk should be weighed? The prospective babies, however, have no voice in this assessment—to what extent should the government speak for them versus their potential parents? These questions will not be easy to answer, but these kinds of questions are not easy in many cases of drug, biological product, or medical device approvals. Yet they get made.

He Jiankui did not go through any of the difficult and painstaking steps set out above for deciding whether to attempt human germline genome editing. He claims, with no publications, that he did some preclinical trials in some nonhuman animals—with no details. But he made the decision to go forward with his clinical trial alone, without any outside regulatory oversight, except, perhaps, from a local hospital's ethics committee. And that, to me, was his greatest sin—he acted recklessly, with inadequate evidence for the safety of his procedure for the children he was trying to produce, and with no transparency or oversight.

Even if an intervention has been approved by FDA, its safety is not set in stone once it has been approved, based on clinical trials. Its actual clinical use is likely to be much bigger than the clinical trials and in a much less controlled way. If any method of human germline genome editing is approved as safe and allowed, what should be done to make sure it is used safely? Two different methods should be considered.

The first is continued, and continuous, monitoring of the safety of the children born (and the pregnancies started) using such a method. This would provide a better check on the safety results in the clinical trials, serve as surveillance for any change in safety with changes in implementation, and provide useful data for assessing the safety of facilities that performed human germline genome editing. Clinical practice may show different safety issues for many reasons. First, the numbers will be much bigger. If a particular safety problem occurs in 2 percent of cases, a trial with 50 births may well not show it at all. Bigger numbers can probe for less common problems. Clinical practice also often involves patients who are different from those recruited for, and accepted into, clinical trials, as well as physicians who may not be as well-trained or competent as those in the trial. And, depending on both any legal restrictions on use—and how any such restrictions are defined and applied—it may well involve patients who are using the approach to make changes that were not studied in the clinical trial. Similarly, the safety of intentional changes, such as a change in the composition of the culture medium or the time at which the embryos are edited, could be monitored. And, of course, results from individual clinics could be used to assess the safety of those clinics.

That leads directly to the second approach—limiting what facilities can provide human germline genome editing. Currently, the United States has no national licensure or approval of fertility clinics; neither do any states have special licensing requirements for those clinics. To the extent the clinics function as clinical laboratories, they are subject to some federal regulation under the Clinical Laboratory Improvement Amendments Act[23] as well as accreditation by the College of American Pathologists. According to the American Society for Reproductive Medicine (ASRM),

ASRM and the College of American Pathologists administer a reproductive laboratory accreditation program for embryology labs to assure that they conform to high national standards of quality. ASRM also produces ethics and practice guidelines. Its affiliate, the Society for Assisted Reproductive Technology (SART), strictly monitors member clinics for adherence to ASRM guidelines, accreditation of their embryology labs, qualification of their staff, and submission of data to the CDC.[24]

All the physicians at such facilities must be licensed by, and can be disciplined by, their state medical boards. The specialty physicians, mainly ob/gyns and some urologists, will normally be certified by their respective specialty colleges.

But, in the United States, no one has the mission to say to a clinic, "Yes, you can practice fertility medicine," "You can no longer practice fertility medicine," or "You are competent to practice these procedures but not those." The United Kingdom does have such a system, administered by HFEA. Human germline genome editing is only one of many assisted reproduction technologies, and, if I am right, it is unlikely to be very important, at least for several decades. It might provide a good spark for increasing the regulation of clinic-by-clinic, procedure-by-procedure safety—if such regulation were to prove politically feasible.

15   The Big Decisions—And How to Make Them

The last chapter dealt with assessing (and potentially assuring) the safety of human germline genome editing and ultimately making decisions based on safety-related evidence. But safety is not the only criterion, and may not be the most important. Proof of reasonable safety should be essential to any approved clinical use, but it may well not be sufficient. Two other big decisions that go beyond just scientific and medical questions must be answered: Should human germline genome editing be banned altogether, and, if not, should it be permitted only for limited uses and under certain conditions? And perhaps the most important question is how to make those two decisions. This final chapter explores those issues.

Should Human Germline Genome Editing Be Completely Banned?

After the earlier discussion in this book, you should not be surprised to learn that I find no good arguments for an unconditional ban. Such a ban would, at least in strong form, be saying that there are *no* circumstances under which human germline genome editing should be permitted—no amount of proven safety, no limitations on inappropriate use, no provisions for just access, no prospective parents whose need to use the procedure to have healthy genetic children would justify it. I think this argument cannot be based directly on the consequences of human germline genome editing but instead must be founded on the belief that such a procedure is inherently wrong. I accept that people may hold such a position, but I do not see how they can hold it consistently with acceptance of a great deal of modern life, modern medicine, and modern reproduction.

At least with regard to reproduction, Catholic doctrine would be consistent in refusing to countenance such a procedure, though the Scriptural and theological arguments against (almost) any kind of interference in the natural processes of human reproduction are, to me at least, obscure. As I understand it, the basic Catholic position on reproduction (apart from any additional issues such as the destruction of embryos) is that it is impermissible to separate the unitive and procreative significance of sex. Why not? Because by "safeguarding both these essential aspects, the unitive and the procreative, the conjugal act preserves in its fullness the sense of true mutual love and its orientation toward man's exalted vocation to parenthood."[1] Artificial insemination and IVF are both condemned by this doctrine. How can one argue about that—probably not with empirical evidence about "the sense of true mutual love"?

Others, often using less argument, will assert from other religious traditions, from philosophical positions, or from a belief in "naturalness" that human germline genome editing is inherently wrong, and wrong not just for themselves but for anyone and everyone at all times and in all circumstances. But in light of the continuity pointed out in chapter 11 between such editing and the many existing ways in which human germline genomes change, and are changed by humans, it is hard for me to see a strong argument in this—one that might convince other people—that does not require throwing out a vast amount of our present civilization and medicine. Some could counter that this is an inevitable problem with situations of incremental changes or variations to declare "this last step is too far." Citing the fabled frog that boiled to death in a pot of slowly heating water, they could say that some line must be drawn somewhere and human germline genome editing is a good place to draw that line. But then they have to justify both that some line is necessary and that this is the place to draw it, in any and all circumstances.

I respect, and would strongly argue that governments should respect, the rights of individuals to decline to use this technology for themselves for whatever reasons, of principle or of caprice, however unconvincing I might find them. But imposing a total ban on this tool's use by other, willing people, in all circumstances, requires some other kind of argument.

At least two other, less absolute, arguments for a total ban exist. One is a strong form of the precautionary principle—that we should not allow such a change unless we are confident that it will have good consequences. I accept the precautionary principle in a weaker form—we should take the time and care to examine the likely consequences of a major change and to have concluded it is not unreasonably risky before implementing it. But I do not accept the strong form. As a general matter, it seems to me to fail to acknowledge that not acting is itself an action, with imperfectly predictable results. A more specific objection in this context is that no strong and unpreventable harms seem likely from this procedure if carefully implemented—in fact, if my assessment is right, it is likely to be rarely used. And general hand-waving about "unpredictability" may argue against uncontrolled and unregulated uses, but, unless taken to an extreme that would, in effect, try (futilely) to ban any changes in anything, probably would not make a convincing case against some carefully controlled and regulated uses. That is especially true if the outcomes were well-monitored, with provisions for examining data on the results and trends and possibly reconsidering the procedure's use at regular intervals.

The other, somewhat more nuanced argument for a total ban focuses not on uncertainty about the results but on a skeptical, perhaps cynical, view of their certainty—that control over the procedure will inevitably be lost, leading to all possible harms from unrestricted use. It assumes we cannot control any change to make sure it will be used for good purposes and that any change will inevitably end up furthering greedy, oppressive, unjust, racist, and other bad ends. This outlook, if taken to its logical end, seems an unattractive invitation to total inaction—if everything, presumably one's own well-intentioned movements, will be misused by the powers that be for evil ends, what is the use of any action?

To me, more convincing is what I see as the empirical evidence that this is just not true—or, at least, not always and inevitably true. I believe technological and social changes can be regulated in ways that avoid the worst results because we have seen that happen in many ways—I believe I have seen it in many ways during my own lifetime, from rights movements for racial minorities, women, LGBTQ people, people with disabilities, and others to modern medicine and health care, the computer revolution, increased travel opportunities, mass and social media, and many other changes in my lifetime. Not a single one of those changes has been implemented in ways that are unequivocally good, and, for some, the net balance of good and bad may still be up in the air (social media is a possible example). But neither have they inevitably collapsed into tools for further oppression. Most directly, fears in assisted reproduction about artificial insemination, fertility drugs, IVF, surrogacy, and other practices have not been realized, at least in full. Why should human germline genome editing be different?

The foregoing paragraphs may just be a rationalization of my general view of the world, and one held widely by others with law training, that the right answer almost always starts with "It depends." But I believe, for the most part (it depends), in that view. If it is accurate, it cautions against any unconditional bans, at least until the possibilities of helpful conditions on the procedure have been carefully explored. And so we move to the second question: Are there conditions or circumstances under which human germline genome editing should be permitted?

Should Human Germline Genome Editing Be Regulated?

Human germline genome editing could be regulated without being banned. Some limits on what it could be used for, in what settings, and by what kinds of intending parents are likely to be suggested, and, in some places, adopted. But there are also questions about who should regulate it, and how.

One might limit the use of human germline genome editing only to specific purposes. I think the likely purposes can be divided into two main classes: disease-related and non-disease-related. The disease-related purposes can best be examined by thinking of autosomal recessive diseases, conditions like sickle cell anemia, beta-thalassemia, or cystic fibrosis, where a child inherits a disease-linked version of the relevant gene from each parent. In those cases, the relevant purposes can be divided into two main groups: avoiding having a child who is or will be affected with the disease and avoiding having a child who, though not affected, has one copy of the disease-linked DNA and so is an unaffected "carrier" of the disease.

With autosomal recessive diseases, a person can have two copies of the disease-linked variation, one disease-linked copy and one healthy copy, or two healthy copies. The first person will have the disease, the second person is healthy but a disease "carrier," while the third is a noncarrier. In these cases, all couples would necessarily fall into one of six categories: diseased/diseased, diseased/carrier, diseased/noncarrier, carrier/carrier, carrier/noncarrier, and noncarrier/noncarrier.

The narrowest and most tempting limitation would be to allow human germline genome editing only for the first category of couples, people all of whose children must have the disease unless germline genome editing is used. There are equivalent situations apart from autosomal recessive diseases. In an autosomal dominant disease, where only one disease-linked copy of the gene is needed to cause the disease but the prospective parent had two disease-linked versions, any child would necessarily receive one of that parent's disease-causing genes and hence would have the disease. Or when a woman has a mitochondrial disease, her mitochondria with the disease-causing DNA variant would always be inherited by all of her children. What unites these cases is that preimplantation genetic diagnosis or prenatal genetic testing could not let the parents have a healthy child. This is the most compelling use of human germline genome editing.

But from there, things get complicated. What if the genetic variations do not always cause the disease, or are not always "fully penetrant"? Would human germline genome editing be allowed if the children who got disease "predisposing" genes only got sick 80 percent of the time? Fifty percent? Twenty percent? What if the disease would not affect *all* of a couple's children but would affect all boys—a disease carried on the Y chromosome, for example, or one on the X chromosome where the mother has two disease-causing copies—or all girls, as where the man's X chromosome bears a dominant disease variation? What if the disease is not particularly serious, such as a mild version of red/green color blindness? What if the disease is treatable after birth, by drugs or by gene therapy? What if the treatments are not always completely effective, or completely safe?

The other disease-related use would go farther and allow parents to use DNA editing to edit embryos not just so they will not have a particular disease but also so they will not be carriers. Let's go back now to autosomal recessive diseases and those six kinds of couples. The second (diseased/carrier) and fourth (carrier/carrier) kind are at risk for having both children with the disease and children who are carriers. The third (disease/noncarrier) category has no risk of having an affected child, but all their children will be carriers. The fifth kind (carrier/noncarrier) are at no risk for having affected children but may have carriers. Only the sixth group (noncarrier/noncarrier) cannot have either an affected child or a carrier (at least in the absence of a new mutation).

PGD can help the second and fourth groups avoid having affected children, and it can help the fourth and fifth groups avoid having children who are carriers. But couples in the second and third groups cannot produce any noncarrier children and so PGD cannot be used to avoid them. What if parents want to avoid not just a child with the disease but the possibility of having grandchildren with that disease? One might allow couples in those groups to use germline genome editing to avoid carrier children (and hence possibly affected grandchildren). Now, of course, their children could act to avoid having affected children themselves by going through the same PGD process when they decide to have children, but editing the disease-linked variations to healthy ones would make that step unnecessary. It would "cleanse" their children's genomes of the unhealthy versions.

So far we've talking about cases where, because of the parents' genes, the children *must* inherit disease-causing or predisposing genes. But what about parents whose DNA variations mean they might have sick children, healthy but carrier children, or noncarrier children. PGD should be able to help those people pick embryos without the disease-causing genetic variations— but only if they have enough embryos and good luck. It could be that the embryos they have made do not include any with their preferred choice. And, for some of them, they may not be able to make any more embryos, probably from a shortage of viable eggs or sperm from the prospective parents. In that case, the prospective parents might be allowed to seek an exception to use editing to change the DNA in one of their already created embryos to turn it from disease bearing to healthy, or, if the jurisdiction accepts avoiding carrier status as a good reason for germline genome editing, from carrier to noncarrier.

And now we come to possible uses of human germline genome editing that are not related to disease. It is tempting to call these "enhancement uses," as that is the category that provokes the most worry, but some of them, including the most technically plausible today, do not produce "enhanced" babies but, rather, babies with non-diseased-related traits that their parents prefer—sex, eye color, hair color, skin color. What they all share is the fact of not being related to preventing diseases or disabling conditions, which in turns means that they are, in most places, going to have much less ethical and political support. A country might regulate, probably through banning, any or all such non-disease-related uses of human germline genome editing.

Note that, as with disease-related traits, PGD would be able to provide parents with what they want in some cases but not all. Two parents with light colored eyes probably could not use PGD to select an embryo that would become a child with dark eyes because the parents will not carry the "dark-eyed" DNA variations. They would have to use editing to get that result.

So, in summary, countries might allow, or ban, using genome editing for the following:

1. to avoid having a child with a genetically inevitable disease,

2. to avoid having a child with a genetically inevitable carrier status,

3. to avoid having a child with a genetically inevitable disfavored nondisease trait, and,

4. in each case, countries might allow, or ban, using the method to avoid any of the three categories where the problem is not genetic inevitability but a lack of existing and likely future embryos with the desired variations.

Each of those categories, of course, will be plagued by the uncertainties noted above about the likelihood of the disease or trait, its seriousness or importance, or to what extent it could be treated or achieved by methods other than human germline genome editing.

One issue with this kind of regulation of types of uses of human germline genome editing may appear to be only semantic, but could, in practice, be important. One could either frame the regulation of certain uses as permitting some uses or as forbidding others. This could be important because it may well set the burden of proof. If one permits only specific uses, any use that is ambiguous or unclear would have to prove that it, in fact, fit inside the allowed scope. If one prohibits only specific uses, someone would have to prove that an ambiguous or unclear use fit inside the definition of prohibited uses. Does uncertainty or ambiguity mean that the use can go forward or not? One could spell that out explicitly, but the choice of specifying permitted uses versus banned uses sets the framework implicitly.

Where should human germline genome editing take place? The first answer to that question is in settings where it can be provided safely and effectively. We should not encourage, or allow, "do-it-yourself" embryo editing or genome editing by, say, chiropractors or naturopaths. For safety reasons, only licensed physicians and clinics should be allowed to use the techniques. If a jurisdiction imposes limits on the procedure's use, requiring that only licensed professionals provide it offers other benefits—it limits the number of providers who have to be monitored for compliance with those rules. It also probably means that those providers will have some valuable assets at risk if they violate any rules, such as their licenses.

But is it enough to say that only licensed physicians and clinics can do human germline genome editing? Even assuming this would mean, either in law or in fact, only fertility clinics, the United States has about 500 clinics. Under existing U.S. law, once a drug, biological product, or medical device has been approved, normally any physician can use it for any purpose he or she wants. If we think human germline genome editing is especially risky or tricky, or just especially important, we may want only to allow it to be done in a small number of clinics, and by a small number of professionals, who have shown that they are—or are likely to be—particularly good at the procedure. The HFEA does this in the United Kingdom. Each clinic has to be separately approved for each use it makes of assisted fertility technologies—not just for PGD, or intracytoplasmic sperm injection, or any other technique, but, at least for PGD, for each and every disease or trait that clinic wants to select against, or select, using PGD.[2] A clinic in Newcastle, for example, would have to get specific permission before it can do PGD to detect Huntington's disease.

No mechanism exists for this kind of detailed regulation in the United States. It could be created, but it would likely run into opposition from clinics that feared losing business opportunities, as well as from the American Medical Association and other doctors' groups, which fear any interference with "the practice of medicine."

The FDA does have a mechanism called "Risk Evaluation and Mitigation Strategies" (REMS) under which it can sometimes limit what conditions physicians can prescribe drugs or biological products for—and which physicians can prescribe them. FDA gives as an example a REMS for an antipsychotic drug that sometimes causes a serious reaction called "post-injection delirium sedation syndrome." FDA required the drug's manufacturer to develop a REMS to "ensure that the drug is administered only in certified health care facilities that can observe patients for at least three hours and provide the medical care necessary in case of an adverse event."[3] Perhaps FDA could subject germline genome editing to such a REMS, though it is noteworthy that the REMS itself must be developed by the manufacturer and then approved by FDA. Even if FDA were to order a REMS, who would be the "manufacturer" for germline genome editing? (We will come back to that question and others around FDA.)

There is another important issue about the setting: Should a jurisdiction try to prevent its nationals from going to another jurisdiction to get human germline genome editing that it does not allow? This issue of so-called "reproductive tourism," a subset of the broader "medical tourism," is not new.[4] It potentially applies not only to different rules between different countries but different rules between different states. Countries mainly regulate activities within their own borders but sometimes countries do regulate, even to the point of criminal sanctions, what their nationals do somewhere else. In the United States, for example, for an American national to have sex with a child is a federal crime even if it is committed in a country where it is legal.[5]

Sometimes agreements between jurisdictions permit this kind of medical travel, as in the European Union, where any E.U. citizen has a right to travel to any E.U. country for medical treatment.[6] Although I know of no court cases of it in the medical context in the United States, I strongly suspect the right to interstate travel would not allow, say, South Dakota to punish one of its citizens for going to California to receive a medical procedure illegal in South Dakota but permitted in California.

Note well, though, that even if one jurisdiction can punish, or try to punish, its nationals from getting a particular medical treatment elsewhere, enforcing that could be very difficult. How will customs officials at an airport know whether the returning citizen is coming back carrying a two-week-old implanted, gene-edited, embryo? Even if they could detect the pregnancy, how would they know whether it was the result of sex, an IVF clinic, or an IVF clinic that had done germline editing?

Should the prospective parents allowed to use human germ-line genome editing be restricted? This category contains at least two different kinds of regulation, one of which might not be recognized as regulation though it acts that way.

First, some jurisdictions may decide to limit what kinds of people can become parents of germline genome edited children. Some may decide that they must be married, or must be "solidly" partnered, or must be heterosexual, or must be over or under a certain age. These kinds of limitations may seem odd to many of you, but every one of them exists with respect to IVF in some places. In some Muslim countries only married (which there means heterosexual) couples are allowed to use IVF.[7] France makes IVF illegal except when used by a married or solidly partnered heterosexual couple.[8] Italy does the same as France; on top of which, in Italy women who are not "of fertile age" may not use IVF to try to have a child.[9] Other countries may have limitations that do not directly restrict the legality of germline genome edited children but do so indirectly. In jurisdictions that ban surrogacy, for example, gay couples could not have their own genetic children at all, including through germline genome editing. Whatever one thinks of these limitations, they would, without modification, almost certainly apply to human germline genome editing as it will (almost) always require IVF.[10] On the other hand, it is not clear to me that anyone would want to impose specific limitations on which prospective parents could use germline genome editing, beyond those genetic limitations that could come from limiting the technique to people with particular genetic risks.

Thus far I have talked of express, legal limitations on who is eligible to use this technology. But many times the limitations are not prohibitions but financial concerns. Judith Daar recently published a book called *The New Eugenics*, arguing forcefully that the current American system of paying for assisted reproduction effectively is eugenics against the poor.[11] In the United States assisted reproduction is rarely covered by health insurance and never by Medicaid, the program that pays for health care for the poor—and for about half of all pregnancies and childbirths in the United States.[12] If IVF is expensive, only those with money get IVF. Even in countries where IVF is covered by public funding, the health plans restrict who can use IVF and how many times, which, de facto, work as limits on who can have IVF.

Germline genome editing is likely to be expensive. A cycle of bare-bones IVF in the United States currently costs between $10,000 and $15,000, with various attractive bells and whistles that quickly and strongly boost that price tag.[13] Adding the cost of the creation and use of the CRISPR construct, and the added monitoring that genome editing would necessarily entail (did the right places get edited? did any "wrong places" get edited), the process will not, unless things change radically, be cheap. And even in countries that cover some IVF, germline genome editing may well not be quickly accepted as covered, both for its controversial nature and because it would make the systems more expensive.

Should countries make sure that all their nationals can use human germline genome editing? Perhaps, but will they? And, if so, will covered editing be more restricted than what is legal? As a parallel, the British National Health Service will cover IVF for women up to a certain age; beyond that, the government system will not pay for it, but it is not illegal.[14] Older prospective parents can get IVF in the United Kingdom—if they have enough money. A health system might pay for it for couples where both members carry the same autosomal recessive condition and allow it for some other uses—allow it, but not pay for it. Who can and cannot use this technique, because of either legal prohibitions or economic constraints, has important ethical implications.

Those implications do not stop at national borders. It could be that rich countries will cover human germline genome editing for some or even many prospective parents. But it will be a long time before Burkina Faso, Laos, Haiti, or other poor countries can do the same. Economic unfairness may be limited on a national scale; limiting it internationally will not happen quickly, at least not without massive subsidies that seem unlikely to appear.

A final issue involves implementation more broadly. Who, if anyone, will govern the process and determine that a couple's planned use does or does not qualify as permitted (or banned)? Should it be professional self-regulation, regulation through an administrative agency (perhaps something akin to the United Kingdom's HFEA), decisions by courts in judicial actions (civil or criminal), or something else?

That issue is particularly pressing in the United States. The FDA has asserted jurisdiction over human reproductive cloning, human mitochondrial transfer techniques, and human germline genome editing on the ground that the "more than minimally manipulated" human cells involved are drugs or biological products. (The differences between these two regulatory categories are not important in this context.) Whether FDA is interpreting its jurisdiction properly in these cases is somewhat controversial.[15] On its face, it does seem "odd," at least, to consider a human embryo as a drug or biological product. I think the claim would probably be upheld in court, in part because of the breadth of the statutory language defining drugs and biological products, but also because otherwise human cloning and mitochondrial transfer would have effectively no regulation in the

United States. The situation is less dire with regard to germline editing, where FDA also takes the much more intuitive position that the CRISPR construct used for such editing is a drug or biological product.

Assume human germline genome editing is within FDA's jurisdiction and so falls into the FDA's usual requirements for approval: preclinical trials followed by clinical trials under an IND to prove enough safety and efficacy to allow eventual approval (a New Drug Approval for drugs or a Biological License for biological products). So what's the problem? Well, human germline genome editing does not seem to fit well into other aspects of the FDA's scheme. For one thing, FDA requires a sponsor to apply for the IND and the eventual approval of a New Drug Application or Biological License Application. That's usually easy—it is the company that plans to market it, a firm that almost always has a patent position that it can use to prevent others from using the compound. Who would be the sponsor for human germline genome editing? And who, if anyone, will eventually have patents that cover the process?

Firms seek approval for drugs and biologicals for a particular indication—say, a particular chemotherapy drug for treating gastrointestinal stromal tumors, which is the indication that will have been the subject of the clinical trials. Would sponsors, whoever they are, have to get separate FDA approvals, based on proof of safety and efficacy, for using germline gene editing to prevent cystic fibrosis and sickle cell disease and beta-thalassemia? But federal law also allows "off-label use," where any doctor can use any approved drug for any purpose, whatever it was approved for. The sponsor of the drug can legally sell it to people knowing that they will make an unapproved use of it, but is, under current law, quite limited in how it can promote those off-label uses. Will the first sponsor to get FDA approval for any human germline genome editing be able to sell it (whatever that means—license it for use by fertility clinics?) for all human germline genome editing?

Last, FDA approvals usually come with special benefits, intended to encourage development of new treatments. Some involve exclusivity periods. An approved drug that involves a "new chemical entity" is entitled to five years during which FDA promises not to approve a competing drug for the approved use. An approved biological product gets 12 years of exclusivity. A sponsor can seek designation for its product as an "orphan drug" or "orphan biological" as long as the estimated number of people in the United States with the relevant indication is fewer than 200,000. Approved orphan drugs get their own seven years of exclusivity (though it runs concurrently with the other exclusivities).[16] Would those rules apply here?

The law also provides "priority vouchers" for treatments for some conditions—currently including certain tropical diseases, rare pediatric conditions, flaviviruses (of which Ebola is the most famous), and biological warfare countermeasures. The sponsor may use or sell the priority voucher to speed FDA consideration of a different application, even one that has nothing to do with a rare disease. It usually leads to an FDA decision four months earlier, and hence, if the application is approved, sales starting four months earlier. Priority vouchers have been resold for as much as $350 million (though most sales recently have been below $100 million).[17] Germline genome editing could certainly be used to treat some of the stated conditions—would that generate a priority voucher?

In short, even apart from the deeper questions of FDA versus HFEA versus a judicial or other model, the FDA process seems to fit poorly something as broad and as basic as human germline genome editing.

How Should We Decide?

And, last, we come to the metaquestion—how to answer the other questions. Jurisdictions—mainly nations but in some cases subnational jurisdictions (states) or supranational jurisdictions (the European Union or the Council of Europe) will have to decide whether to allow human germline genome editing at all and, if so, with what safety precautions and limits on uses, providers, and prospective parents. What kind of decision-making process should we seek?

This book has already discussed various positions taken about the role of the public in governing human germline genome editing, but it may be useful to review them here. The statement of the organizing committee of the first human genome editing summit said, "It would be irresponsible to proceed with any clinical use of germline editing unless and until . . . (ii) there is broad societal consensus about the appropriateness of the proposed application."[18] The 2017 National Academies report on human genome editing said that even clinical trials for disease prevention should proceed only if they met all of 10 requirements, including "continued reassessment of both health and societal benefits and risks, with broad ongoing participation and input by the public . . ."[19] Others have called for human germline genome editing to go forward only in the context of a broad international consensus.[20] In March 2019 a group of scientists called for a formal moratorium on human germline genome editing for a fixed period of years but then suggested a more limited demand for consensus. After the moratorium period, any nation could also choose to allow specific applications of germ-line editing, provided that it first: gives public notice of its intention to consider the application and engages for a defined period in international consultation about the wisdom of doing so; determines through transparent evaluation that the application is justified; *and ascertains that there is broad societal consensus in the nation about the appropriateness of the application.* Nations might well choose different paths, but they would agree to proceed openly and with due respect to the opinions of humankind on an issue that will ultimately affect the entire species. [Emphasis added.][21]

The presidents of the U.S. National Academies of Sciences and of Medicine and of the U.K. Royal Society responded in an editorial, arguing against a moratorium but stating that

we also recognize that—beyond the scientific and medical communities—we must achieve broad societal consensus before making any decisions, given the global implications of heritable genome editing.[22]

So we know that people want a consensus, or public participation and input, or both, but what does that mean, and what should we do? Like (almost?) everyone who has commented on the issue, I think this is not a decision for scientists alone to make, either singly, as in the He Jiankui case, or collectively, through various scientific academies. The public is interested in, and necessarily involved in, this kind of issue, with its potential implications—strong or weak, near term or long term—for the future of our species. As I argued above, Science should not try to be separate and apart from its societies—and, as a practical matter, it cannot be, no matter how hard it tries.

On the other hand, a global consensus is a chimera. Seven and a half billion people are not going to agree on this issue— nor will the fraction of those who understand it. Neither will the roughly 200 countries of the world, at least at anything other than a lowest common denominator. Not all countries have agreed to various nuclear weapons or climate change treaties in spite of the apparently obvious need for them. If somehow something close to unanimity were achieved, it would probably be at the cost of precision. Thus, the Council of Europe enacted a convention that banned human cloning, but effectively left open the then-hot question whether it covered just reproductive cloning or "research cloning" (of human embryos for only ex vivo use) by not defining a "human being."[23] It also left the implantation and enforcement of such a ban up to the member nations, some of whom were probably happy, for political reasons, to sign it but will have little interest in enforcing it.

Like (almost?) everyone else, I think the public needs to play an important role in these decisions. Perhaps my legal training is showing too much, but I think the proper way for that role to play out is through the legal system in each relevant jurisdiction. Human germline genome editing should be allowed, or banned, or limited, by governments. As I am skeptical that any functional international treaty will be adopted (and enforced), that leaves me largely with nations, or, in some cases, subnational or supranational bodies. Human germline genome editing should not proceed, even if shown safe and effective, without a role for public opinion at the relevant level (usually national).

That role, however, should not be set to require a consensus, a mushy concept at best. It should be expressed through the jurisdiction's procedures. In some places, such as Switzerland or California, that might mean legislation through a referendum. I do not, however, think a referendum should be required in all jurisdictions. Probably because I do not think human germline genome editing is likely to change our world or our species in extraordinary ways—at least, not unless or until major advances are made over decades in our understanding of genetics—I do not think extraordinary signs of approval or agreement are necessary. If a jurisdiction enacts new legislation authorizing, with whatever limitations, this kind of work, that should be enough to justify it, just as it justifies other government actions or approvals, in almost all circumstances.

It does not justify government actions that breach that jurisdiction's constitution or international treaties it has entered into. Some would argue that legislation is also invalid if it conflicts with universal human rights, though how those are defined, and by whom, are sticking points. Those exceptions seem irrelevant here. Few, if any, will argue there is a constitutional or universal human right either to use human germline genome editing or to prevent anyone from using it. (I say "few" confidently—I am confident that some will, but I am also strongly suspect their arguments will not be widely convincing.)

But what about something less than approval as expressed through new legislation? Is the public willingness criterion met not just by new legislation or law but by older laws that cover this technology? In the United States, for example, that would include the federal Food Drug and Cosmetic Act and the Public Health Service Act, under both of which FDA asserts jurisdiction over this kind of intervention. I think continued acquiescence in existing laws should be enough. If enough of the public opposes human germline genome editing, it should have a good chance—in most countries, at least—of bringing political pressure to force action to restrict it. Conversely, where the existing law bans human germline genome editing, as it does in most of Europe, that should count as public disapproval of the practice, unless or until changed.

I say this without believing that political structures accurately reflect "the will of the people" or even the will of the people who fully understood the question. I do think that the political structures in most countries are the best opportunity we have to find a legitimate, if not perfect, assessment of a broader public view—not necessarily a full-throated "love" or "hate" but often a distracted "fine" or "I don't care." We make our laws in most countries through such a method of putative representatives of the people expressing "their wishes"—with formal (elections) or informal (strikes, riots, or revolutions) ways for the people to "correct" their representatives. Human germline genome editing does not seem to me to be important enough to require substantial changes to existing political structures.

So jurisdictions, not international organizations and/or inchoate and unmeasurable consensuses—or, for that matter, individual scientists or doctors or organizations of scientists or doctors—will determine how we answer the big questions about human germline genome editing. They will decide, either directly or through delegations of authority to administrators or administrative agencies, whether to ban it entirely or to regulate it in various ways. Scientists, physicians, patients, and their groups may influence those governments. So may groups strongly opposed to any germline editing, whether religious or secular. But at the end of the day, governments will make the call.

I say this with full awareness of the weaknesses of political structures, even in many democratic countries. But I see no need for, and no plausible good alternative to, national laws and national political structures. Some of those structures are, to my eye, quite bad, and even those that I consider to be presumptively good—liberal democracies—are far from perfect. But ultimately I agree with Winston Churchill:

> Many forms of Government have been tried, and will be tried in this world of sin and woe. No one pretends that democracy is perfect or all-wise. Indeed it has been said that democracy is the worst form of Government except for all those other forms that have been tried from time to time.[24]

In addition to scientific and medical agreement that the procedure is safe, human germline genome editing should not be allowed unless a country's laws, newly created or preexisting, permit it.

Conclusion

So at the end of the saga of He Jiankui and his experiment—
and of the broader implications of human germline genome
editing—what have we learned? Perhaps I should make that
a nonrhetorical question and ask each of you what *you* have
learned. But even in this interactive age, that's difficult to do in
a book. So I'll offer my thoughts on what it all means.

First, I hope you found the story, at least in parts, a fascinating tale, both of events and of some of the people caught up in
those events. And, if you did, perhaps this book did not need to
try to offer any lessons.

But it did. On the one hand, this has been a cautionary tale
about science and scientists. People can overreach. He Jiankui,
driven as far as I can tell by what Macbeth called "vaulting
ambition, which o'erleaps itself,"[1] and aided by a serious lack
of scruples, behaved terribly. He put three lives at needless and
reckless risk (and tried to do the same to more). Ben Hurlbut is
right that he is not a complete aberration but is the product,
to some extent, of some of the pressures, and opportunities, of
today's science. But he is an outlier (as well as an out-and-out
liar)—the extreme case and far from the much more mundane
norm among scientists.

The story is also a cautionary tale *for* Science. He Jiankui damaged Science by reinforcing the Victor Frankenstein image—the mad, uncontrolled scientist. Most scientists, and hence most science, are much more rule bound, constrained by the needs of getting and keeping jobs, tenure, and most importantly grants, as well as wrapped, in most countries, in many bureaucratic threads of control. But He happened, and Science needs both to act to minimize the harm he has caused and to be seen to be doing so. It cannot exist apart from its societies; it requires their support, or, at the very least, their acquiescence. Cases like He's mean that Science will have to work harder to gain and keep those.

And it is a story about an exciting and frightening new biological technology. Human germline genome editing is not the first such promising and controversial biological technology—remember IVF, cloning, embryonic stem cells, and others. But, as has been the case before, a clear hard look at the technology shows that it is neither as good nor as bad as the first impressions it has made, impressions aided by media, scientists, university press offices, columnists, novelists, perhaps even law professors and bioethicists, and others eager to grab our attention.

To quote my conclusion to an earlier article on this topic.

> But, like Dolly's birth, He Jiankui's CRISPR'd babies are not the end of the world—or the beginning of the end of our species. They are a challenge both to the ability of Science to regulate itself, and to the world's trust in Science. Drastic action is not needed but, in its aftermath, useful things should be done. And, just as important, useful things must be *said*.[2]

Acknowledgments

Should I start by acknowledging Dr. He Jiankui? Without him, this book would have been neither possible nor necessary, though the world, and the world of science, would have been a better place.

I'd rather start with an old friend, Dan Clendenin, who sent me an email at 7:37 p.m., Sunday, November 25, 2018, whose subject line—"CRISPR Babies"—was my first clue about what was going on. He, and many others, had heard me talk about gene-edited babies but always as something fairly distant. He got to be the first one to point out to me that my time predictions had been wrong.

This did not start out to be a book, but, over the next few days, weeks, and months, I found myself spending many hours reading about, listening to, and thinking on He's experiment. That gradually became many hours giving talks about it and then writing about it. After November 2018 and before finishing this book, I published four articles, ranging from over 30,000 words to under 1,500, about various aspects of the He Jiankui case, plus an October 2017 piece that helped prepare my thinking for these events. I want to thank all the journals involved, *The CRISPR Journal*, *The Journal of Law and the Biosciences*, *Issues in Science and Technology, STAT News,* and *Leapsmag,* as well as the editors and staff I dealt with—Kevin Davies and Rodolphe Barrangou, Glenn Cohen and Nita Farahany, Kevin Finneran, Sharon Begley and Patrick Skerrett, and Kira Peikoff.

By the time I had written that much on the topic, I seemed to be well on my way to a book. I also was worried that, if I didn't write this up as a book, it would take over my planned next book (tentative title, Playing with Life), which I wanted to be about human modifications of nonhuman organisms. At this point, my MIT Press editor, Robert Prior, stepped in, and we started talking about a book on this topic. I owe him thanks for his confidence in the project (and me), for his advice, and (of course) for the contract.

For crucial help in turning this all into a book, though, I have to thank the then–Stanford Law School students who worked on it as research assistants: James Rathmell, J.D. and MBA 2019, and Brittany Cazakoff and Christie Corn, both J.D. 2020. This book was finished in my 35th year of teaching at Stanford Law School. I have been lucky to have had many great student research assistants during those years. Some may have been as good as James, Brittany, and Christie and their rare combination of legal and biological sophistication and great work ethics, but none has been better. Throughout the process, Stanford's extraordinary Robert Crown Law Library, under the direction of Beth Williams, provided everything I needed, often before I knew I needed it.

Which leads me a step farther back in my thanks, to before this book was even a twinkle in my eye. In 1990 my Stanford colleagues Paul Berg, David Botstein, and Lucy Shapiro added me to the planning committee for a two-and-a-half-day Stanford Centennial Symposium on the then-new Human Genome Project, held in January 1991. That launched me on the professional path of my last 30 years. Over the ensuing 30 years plus, scores of scientists, including some with significant roles in this book, have been willing to spend their time explaining things to a law professor whose last biology class was in tenth grade and to do so with patience, grace, and friendship.

This is an unusual effort for me because I personally know many of the people involved. I have spoken or corresponded with some of them about human germline genome editing in general and the He experiment in particular. I want to note for special thanks Jennifer Doudna, Alta Charo, David Baltimore, Paul Berg, George Daley, and the other participants in the January 2015 Napa meeting, as well as Anne-Marie Mazza, Matt Porteus, Steve Quake, Bill Hurlbut, and Ben Hurlbut. I particularly want to thank Porteus and Bill Hurlbut for taking part in a panel I organized at Stanford Law School in January 2019.

For this book, which lies uneasily somewhere between history and journalism, I also want to thank the many journalists who talked with me. They were, for the most part, looking to me as a source, both for background and for quotes, but they served as a source for up-to-date information for me, both in the conversations I had with some of them and in the remarkable reporting they all produced. And I also want to thank the reporters I did not talk with, whose work provided much of my knowledge of the subject, as the references to this book indicate. I know I will forget some (and beg forgiveness), but, at the very least, the combined list includes Antonio Regalado (whose investigative reporting broke open the story), Jane Qiu, Sharon Begley, Andrew Joseph, Sara Reardon, David Cyranoski, Preetika Rana, Ed Yong, Carl Zimmer, Carolyn Johnson, Christina Farr, Kristin Brown, Megan Molteni, Malcolm Ritter, Jon Cohen, Pam Belluck, Julia Belluz, Lisa Krieger, and Marilynn Marchione.

I need to acknowledge one other, probably unusual, source of help as I explored this topic: Twitter. I hear horror stories about Twitter and, on occasion, stumble into corners of it where those stories play out, but the loose grouping that I think of as "bio-twitter," with its mix of scientists, journalists, analysts, ethicists, and just plain interested people, is uniformly polite and informative, as well as (usually) fun. I learn a lot from them.

Finally, last but first, my beloved wife, Laura Butcher. Without her encouragement and support—and willingness to lose me to my computer for many evenings and nights—this book would never have been written.

Notes

Introduction

1. Antonio Regalado, "Exclusive: Chinese Scientists Are Creating CRISPR Babies," *MIT Technology Review*, November 25, 2018, https://www.technologyreview.com/s/612458/exclusive-chinese-scientists-are-creating-crispr-babies.

2. "He" is the scientist's family name while "Jiankui" is his personal name—his "first name" in Western usage. Sometimes his name is rendered in the West as Jiankui He—though he encouraged people to call him "J.K.," probably recognizing how hard the Chinese pronunciation of his name is for most English speakers. In this book I will adhere, for He and other Chinese people, to the Chinese custom of putting the family name first.

3. Marilynn Marchione, "Chinese Researcher Claims First Gene-Edited Babies," Associated Press, 2018, https://www.apnews.com/4997bb7aa36c45449b488e19ac83e86d.

4. See, e.g., The He Lab, "About Lulu and Nana: Twin Girls Born Healthy after Gene Surgery as Single-Cell Embryos," 2018, https://www.youtube.com/watch?v=th0vnOmFltc.

5. I. Wilmut, A. E. Schnieke, J. McWhir, A. J. Kind, and K.H.S. Campbell, "Viable Offspring Derived from Fetal and Adult Mammalian Cells, *Nature*, 1997, https://doi.org/10.1038/385810a0.

6. Hiram Caton, "Selling Dolly: An Ethics Hoax," *Bioethics Research Notes* 10, no. 2 (June 1998): 1–7, https://hiram-caton.com/pages/dolly .php.

7. Robin McKie, "Scientists Clone Adult Sheep," *The Observer*, February 23, 1997, https://www.theguardian.com/uk/1997/feb/23/robinmckie .theobserver. McKie subsequently said that he did not violate the embargo but got the information about the lamb from a scientist not directly involved in the research. R. C. Fox and J. P. Swazey, *Observing Bioethics* (Oxford: Oxford University Press, 2008).

Chapter 1

1. Jinzhou Qin, Yangran Chen, Shuo Song, et al., "Birth of Twins after Genome Editing for HIV Resistance," unpublished manuscript, on file with the author.

2. Antonio Regalado, "China's CRISPR Babies: Read Exclusive Excerpts from the Unseen Original Research," *MIT Technology Review*, December 2019, https://www.technologyreview.com/s/614764/chinas-crispr-babies -read-exclusive-excerpts-he-jiankui-paper.

3. Yu Pengfei, "The Lifetrack of He Jiankui: Who Gave Him the Courage," *Sina*, November 27, 2018, http://news.sina.com.cn/c/2018-11-27/ doc-ihpevhck8869227.shtml. This is an article in a Chinese publication. I am relying on Google Translate to read it.

4. Loudi is a "prefecture-level city" in China, which makes it more like a county in the United States. In 2010 (after He left it), it had a population of nearly 3.8 million people and covered over 3,000 square miles. It includes one district, two cities, and two counties. Xinhua County is its largest subdivision, covering about 2,000 square miles and, in 2010, holding over 1.1 million people.

5. Mike Williams, "He's on a Hot Streak," *Rice News Media and Relations*, November 17, 2010, http://news.rice.edu/2010/11/17/new-way-of -predicting-dominant-seasonal-flu-strain.

6. Jiankui He, "Spontaneous Emergence of Hierarchy in Biological Systems," Rice University, 2010, https://scholarship.rice.edu/bitstream/handle/1911/70257/HeJ.pdf.

7. Here's the whole abstract:

Hierarchy is widely observed in biological systems. In this thesis, evidence from nature is presented to show that protein interactions have became [sic] increasingly modular as evolution has proceeded over the last four billion years. The evolution of animal body plan development is considered. Results show the genes that determine the phylum and superphylum characters evolve slowly, while those genes that determine classes, families, and speciation evolve more rapidly. This result furnishes support to the hypothesis that the hierarchical structure of developmental regulatory networks provides an organizing structure that guides the evolution of aspects of the body plan. Next, the world trade network is treated as an evolving system. The theory of modularity predicts that the trade network is more sensitive to recessionary shocks and recovers more slowly from them now than it did 40 years ago, due to structural changes in the world trade network induced by globalization. Economic data show that recession-induced change to the world trade network leads to an increased hierarchical structure of the global trade network for a few years after the recession. In the study of influenza virus evolution, an approach for early detection of new dominant strains is presented. This method is shown to be able to identify a cluster around an incipient dominant strain before it becomes dominant. Recently, CRISPR has been suggested to provide adaptive immune response to bacteria. A population dynamics model is proposed that explains the biological observation that the leader-proximal end of CRISPR is more diversified and the leader-distal end of CRISPR is less diversified. Finally, the creation of diversity of antibody repertoire is investigated. It is commonly believed that a heavy chain is generated by randomly combining V, D and J gene segments. However, using high throughput sequence data in this study, the naive VDJ repertoire is shown to be strongly correlated between individuals, which suggest VDJ recombination involves regulated mechanisms.

8. Shenzhen is a "subprovincial city" in Guangdong Province in Southern China, bordering Hong Kong. It currently has a population officially counted at about 13 million but thought to be, in fact, closer to 20 million. In 1980, when it was made a "Special Economic Zone," its population was 30,000. It has become an industrial powerhouse in the intervening 40 years. https://en.wikipedia.org/wiki/Shenzhen.

9. I tried to calculate the actual distance between the university and Hong Kong on Google Maps, but I got a notice that "Sorry, we could not

calculate driving directions from 'Hong Kong' to 'Southern University of Science and Technology, 1088 Xueyuan Ave, Nanshan, Shenzhen, Guangdong, China, 518055.'" I presume this is because of the continuing disputes between China and Google, but it is a useful reminder of ways in which China is different.

10. Crunchbase, "Direct Genomics," 2019, https://www.crunchbase .com/organization/direct-genomics#section-overview.

11. Crunchbase.

12. Aaron Krol, "Direct Genomics' New Clinical Sequencer Revives a Forgotten DNA Technology," Bio-IT World, 2015, http://www.bio -itworld.com/2015/10/29/direct-genomics-new-clinical-sequencer-revives -forgotten-dna-technology.html.

13. "China's Direct Genomics Demonstrates Single-Molecule Sequencing of E. Coli," Genomeweb, 2017, https://www.genomeweb.com/ sequencing/chinas-direct-genomics-demonstrates-single-molecule -sequencing-e-coli#.XcyYE5NKiHo; Luyang Zhao, Liwei Deng, Gailing Li, et al., "Resequencing the *Escherichia coli* Genome by GenoCare Single Molecule Sequencing Platform," *BioRxiv*, 2017, https://doi.org/ 10.1101/163089. (Deem was the next-to-last author on this paper; He was the senior author.)

14. "Out at His Firm," Genomeweb, 2019, https://www.genomeweb.com/ scan/out-his-firm.

15. Martin Jinek, Krzysztof Chylinski, Ines Fonfara, et al., "A Programmable Dual-RNA-Guided DNA Endonuclease in Adaptive Bacterial Immunity," *Science* 337, no. 6096 (2012): 816–822.

16. Puping Liang, Yanwen Xu, Xiya Zhang, et al., "CRISPR/Cas9-Mediated Gene Editing in Human Tripronuclear Zygotes," *Protein and Cell* 6, no. 5 (2015): 363–372, https://doi.org/10.1007/s13238-015 -0153-5.

17. Jocelyn Kaiser and Dennis Normile, "Chinese Paper on Embryo Engineering Splits Scientific Community," *Science*, April 2015, https:// doi.org/10.1126/science.aab2547.

18. David Cyranoski and Sara Reardon, "Chinese Scientists Genetically Modify Human Embryos," *Nature*, April 2015, https://doi.org/10.1038/nature.2015.17378.

19. Gina Kolata, "Chinese Scientists Edit Genes of Human Embryos, Raising Concern," *New York Times*, April 23, 2015, http://www.nytimes.com/2015/04/24/health/chinese-scientists-edit-genes-of-human-embryos-raising-concerns.html?_r=0.

20. Parts of this chapter, as well as the discussion of the Templeton-funded conference below, draw heavily from an excellent article in *STAT* in December 2018 by Sharon Begley and Andrew Joseph. Sharon Begley and Andrew Joseph, "The CRISPR Shocker: How Genome-Editing Scientist He Jiankui Rose from Obscurity to Stun the World," *STAT*, December 17, 2018, https://www.statnews.com/2018/12/17/crispr-shocker-genome-editing-scientist-he-jiankui.

21. http://blog.sciencenet.cn/home.php?mod=space&uid=514529&do=blog&id=998292.

22. Begley and Joseph, "The CRISPR Shocker."

23. "Genome Engineering: The CRISPR-Cas Revolution," Cold Spring Harbor Laboratory Meetings and Courses Program, https://meetings.cshl.edu/meetings.aspx?meet=CRISPR&year=17.

24. "Genome Engineering."

25. J.K. He, "Jiankui He Talking about Human Genome Editing," 2017, https://www.youtube.com/watch?v=llxNRGMxyCc&feature=youtu.be&t=652.

26. Pam Belluck, "Gene-Edited Babies: What a Chinese Scientist Told an American Mentor," *New York Times*, April 14, 2019, https://www.nytimes.com/2019/04/14/health/gene-editing-babies.html.

27. Sharon Begley, "'CRISPR Babies' Lab Asked U.S. Scientist for Help to Disable Cholesterol Gene in Human Embryos," *STAT*, December 4, 2018, https://www.statnews.com/2018/12/04/crispr-babies-cholesterol-gene-editing.

28. The names of genes and proteins are deeply unintuitive. One useful convention, though, observed in most of the scientific literature, is that the names of genes are *italicized* while those of the protein they provide instructions for—often the same name—are not. If the protein comes from humans and mice, the protein name will also be written in all caps. *PCSK9* is a gene; PCSK9 is the protein whose building that gene directs.

29. Preetika Rana, "How a Chinese Scientist Broke the Rules to Create the First Gene-Edited Babies," *Wall Street Journal*, May 10, 2019, https://www.wsj.com/articles/how-a-chinese-scientist-broke-the-rules-to-create-the-first-gene-edited-babies-11557506697.

30. Paul Knoepfler, "A Conversation with George Church on Genomics and Germline Human Genetic Modification," March 9, 2015, https://ipscell.com/2015/03/georgechurchinterview.

31. Jon Cohen, "The Untold Story of the 'Circle of Trust' behind the World's First Gene-Edited Babies," *Science*, August 1, 2019, https://www.sciencemag.org/news/2019/08/untold-story-circle-trust-behind-world-s-first-gene-edited-babies.

32. Manuscript, on file with author.

33. Cohen, "The Untold Story."

34. Eva Xiao, "China AIDS Group 'Really Regrets' Role in Gene-Editing," Phys.org, 2018, https://phys.org/news/2018-11-china-aids-group-role-gene-editing.html.

35. Sheldon Krimsky, "Letter to the Editor: Ten Ways in Which He Jiankui Violated Ethics," *Nature Biotechnology* 37, no. 1 (2019): 19–20.

36. Preetika Rana, "How a Chinese Scientist Broke the Rules to Create the First Gene-Edited Babies," *Wall Street Journal*, May 10, 2019, https://www.wsj.com/articles/how-a-chinese-scientist-broke-the-rules-to-create-the-first-gene-edited-babies-11557506697.

37. Belluck, "Gene-Edited Babies."

38. Rana, "How a Chinese Scientist Broke the Rules."

39. Manuscript, on file with author.

40. The most detailed analysis of the results I have seen are by Kiran Musunuru in his book *The CRISPR Generation: The Story of the World's First Gene-Edited Babies* (2019, self-published). Musunuru saw both the slides and, eventually, the manuscript.

41. Xinhua, "Guangdong Releases Preliminary Investigation Result of Gene-Edited Babies," *XINHUANET*, January 21, 2019, http://www.xinhuanet.com/english/2019-01/21/c_137762633.htm.

42. Xinhua, "China Focus: Three Jailed in China's 'Gene-Edited Babies' Trial," December 30, 2019, http://www.xinhuanet.com/english/2019-12/30/c_138667350.htm.

Chapter 2

1. This is a nice coinage by my friend Steven Hyman.

2. August Weismann, *Das Keimplasma. Eine Theorie der Verebung* (Jena: Gustav Fischer, 1892.)

3. Trevor Nace, "The World's Largest Organism, Pando, Is Dying," *Forbes*, October 18, 2018, https://www.forbes.com/sites/trevornace/2018/10/18/the-worlds-largest-organism-pando-is-dying/#166f74ff5554.

4. Jacob Sherkow, Patricia Zettler, and Henry Greely, "Is It 'Gene Therapy'?," *Journal of Law and the Biosciences* 5, no. 3 (December 2018), https://academic.oup.com/jlb/advance-article/doi/10.1093/jlb/lsy020/5078563.

5. "Approved Cellular and Gene Therapy Products," U.S. Food and Drug Administration, https://www.fda.gov/vaccines-blood-biologics/cellular-gene-therapy-products/approved-cellular-and-gene-therapy-products.

6. This is similar to, but different from heterozygosity, where one cell will have one edited copy of the gene and one unedited copy. In mosaicism, the editing did not work in every cell. In heterozygosity, it did not work fully in some cells. As noted above, one of the He Jiankui babies was heterozygous, and they both were mosaics.

7. I suspect that is actually how this kind of editing would eventually be done in order to avoid mistakes. If the edited sperm and eggs came not just from the "end product" mature eggs and sperm but from cell lines of the egg and sperm progenitor cells (effectively an edited in vitro germline), those cell lines could be checked to make sure the right edits were made and no off-target edits—or only safe off-target edits—occurred. This could not easily be done in all the cells if an actual embryo were edited. Furthermore, these germ cell lines would also avoid the problem of mosaicism: if clinicians combined the edited egg and sperm to make the fertilized egg, then, barring new mutations, *all* the subsequent cells made by the dividing fertilized egg will have the proper edit.

8. Men continue to make new supplies of sperm from sperm-precursor cells from puberty until their deaths. The consensus (but not unanimous) view is that women, on the other hand, actually make all their eggs before they are born but recruit some to maturity each month during their fertile years.

9. Antonio Regalado, "The DIY Designer Baby Project Funded by Bitcoin," *MIT Technology Review*, February 1, 2019, https://www.technologyreview.com/s/612838/the-transhumanist-diy-designer-baby-funded-with-bitcoin.

Chapter 3

1. A good source for an understandable description of CRISPR and a history of its discovery and development can be found in Jennifer A. Doudna and Samuel H. Sternberg, *A Crack in Creation: Gene Editing and the Unthinkable Power to Control Evolution* (New York: Houghton Mifflin Harcourt, 2017). Doudna is widely accepted as one of the crucial (nonbacterial) inventors of CRISPR; Sternberg was her graduate student.

2. Ian Bogost, "CRISPR Has a Terrible Name," *The Atlantic*, April 2017, https://www.theatlantic.com/science/archive/2017/04/why-does-a-transformative-biotechnology-sound-like-a-cereal-bar/522639.

3. Y. Ishino, H. Shinagawa, K. Makino, et al., "Nucleotide Sequence of the *iap* Gene, Responsible for Alkaline Phosphatase Isoenzyme

Conversion in *Escherichia coli*, and Identification of the Gene Product," *Journal of Bacteriology* 169, no. 12 (1987): 5429–5433, https://doi.org/10.1128/jb.169.12.5429-5433.1987.

4. Francisco J. M. Mojica, Cesar Díez-Villaseñor, Elena Soria, et al., "Biological Significance of a Family of Regularly Spaced Repeats in the Genomes of Archaea, Bacteria and Mitochondria," *Molecular Microbiology* 36, no. 1 (2000): 244–246, https://doi.org/10.1046/j.1365-2958.2000.01838.x. See also Broad Institute, "CRISPR Timeline," 2019, https://www.broadinstitute.org/what-broad/areas-focus/project-spotlight/crispr-timeline.

5. Francisco J. M. Mojica and Francisco Rodriguez-Valera, "The Discovery of CRISPR in Archaea and Bacteria," *FEBS Journal* 283 (2016): 3162–3169, https://doi.org/10.1111/febs.13766.

6. Ruud Jansen, Jan D. A. van Embden, Wim Gaastra, et al., "Identification of a Novel Family of Sequence Repeats among Prokaryotes," *OMICS* 6, no. 1 (2002): 23–33, https://doi.org/10.1089/15362310252780816.

7. Mojica and Rodriguez-Valera, "The Discovery of CRISPR in Archaea and Bacteria."

8. Ruud Jansen, Jan D. A. van Embden, Wim Gaastra, et al., "Identification of Genes That Are Associated with DNA Repeats in Prokaryotes," *Molecular Microbiology* 43, no. 6 (2002): 1565–1575, https://doi.org/10.1046/j.1365-2958.2002.02839.x.

9. Email from Francisco Mojica to author, November 16, 2019.

10. Email from Ruud Jansen to Francisco Mojica, November 21, 2001 (attached to the email from Mojica, November 16, 2019).

11. Email from Ruud Jansen to author, November 18, 2019.

12. Email from Mojica to author, November 16, 2019.

13. Email from Ruud Jansen to author, November 19, 2019.

14. Email from Francisco Mojica to author, November 21, 2019.

15. Jansen email, November 19, 2019.

16. Archaea are the little known third domain of life, along with pro-karyotes (bacteria) and eukaryotes (almost everything else, from amoeba to humans). All archaea are microbes, invisible as individuals, but they are, at the biochemical and genomic level, very different from bacteria. Today, they are often found in very hostile environments, such as hot springs or deep sea vents, but can also be found in less extreme and even cold environments. It is thought that all eukaryotes, including us, are the result of a merger several billion years ago between a bacteria cell and an archaea cell. Laura Eme and W. Ford Doolittle, "Current Biology Archaea," *Current Biology* 25, no. 19 (2015): R851–R855, https://doi.org/10.1016/j.cub.2015.05.025. Laura Eme, Anja Spang, Jonathan Lombard, et al., "Archaea and the Origin of Eukaryotes," *Nature Reviews Microbiology* 15, no. 12 (2017): 711–723, https://doi.org/10.1038/nrmicro.2017.133.

17. Francisco J. M. Mojica, César Díez-Villaseñor, Jesús García-Martínez, et al., "Intervening Sequences of Regularly Spaced Prokaryotic Repeats Derive from Foreign Genetic Elements," *Journal of Molecular Evolution* 60, no. 2 (2005): 174–182, https://doi.org/10.1007/s00239-004-0046-3.

18. It is very debatable whether or not viruses are actually "alive"— they do nothing but inject their own DNA into a victim cell, which then, commanded by the viral DNA, makes more copies of the virus. But, despite being of uncertain "life," we do regularly talk of "killing" viruses, though we could just as easily talk of "inactivating" them.

19. Carl Zimmer, "Breakthrough DNA Editor Born of Bacteria," *Quantamagazine*, February 2015, https://www.quantamagazine.org/crispr-natural-history-in-bacteria-20150206.

20. Frank Hille, Hagen Richter, Shi Pey Wong, et al., "The Biology of CRISPR-Cas: Backward and Forward," *Cell* 172, no. 6 (2018): 1239–1259, https://doi.org/10.1016/j.cell.2017.11.032.

21. See R. Alta Charo and Henry T. Greely, "CRISPR Critters and CRISPR Cracks," *American Journal of Bioethics* 15: 12, 11–17 (2015); Henry T. Greely, "The WorldPost: We Need to Talk about Genetically Modifying Animals," *Washington Post* (December 11, 2017), https://www.washingtonpost.com/news/theworldpost/wp/2017/12/11/gmo

-animals/?utm_term=.f3fc62d8c33a; and Hank Greely, "Why the Panic over 'Designer Babies' Is the Wrong Worry," *LeapsMag* (October 30, 2017), https://leapsmag.com/much-ado-about-nothing-much-crispr-for-human-embryo-editing.

22. Martin Jinek, Krzysztof Chylinski, Ines Fonfara, et al., "A Programmable Dual-RNA–Guided DNA Endonuclease in Adaptive Bacterial Immunity," *Science* 337, no. 6096 (2012): 816–822. If you are the kind of reader who reads the text and notes very carefully, you may have noticed that I said this article was published on June 27, however, the publication date of the article cited is August 17. The formal citations of articles in journals still published in paper is the date (more or less) of the paper publication. But science learns about exciting papers one or a few months earlier, when the online version is posted. A few other citations for this chapter have similar date discrepancies, for the same reason.

23. Jinek et al., "A Programmable Dual-RNA–Guided DNA Endonuclease."

24. Sarah Zhang, "The Battle over Genome Editing Gets Science All Wrong," *Wired*, October 4, 2015, https://www.wired.com/2015/10/battle-genome-editing-gets-science-wrong.

25. Giedrius Gasiunas, Rodolphe Barrangou, Philippe Horvath, et al., "Cas9-CrRNA Ribonucleoprotein Complex Mediates Specific DNA Cleavage for Adaptive Immunity in Bacteria," *Proceedings of the National Academy of Sciences of the United States of America* 109, no. 39 (2012): 15539–15540, https://doi.org/10.1073/pnas.1208507109. The Doudna and Charpentier work did feature an important improvement on what Šikšnys described. Bacteria deploy two different and physically separate kinds of RNA (called a tracrRNA and a crRNA) to use CRISPR; Doudna and Charpentier showed how those two RNAs could be efficiently combined into one construct (the "dual-RNAs" of the last sentence of their abstract); Šikšnys did not describe the tracrRNA.

26. Le Cong, F. Ann Ran, David Cox, et al., "Multiplex Genome Engineering Using CRISPR/Cas Systems," *Science* 339, no. 6121 (2013): 819–824.

27. Prashant Mali, Luhan Yang, Kevin M. Esvelt, et al., "RNA-Guided Human Genome Engineering via Cas9," *Science* 339, no. 6121 (2013): 823–826, https://doi.org/10.1126/science.1232033.

28. Most commonly attributed to Isaac Newton. See Letter from Sir Isaac Newton to Robert Hooke (1675), in *The Correspondence of Isaac Newton: 1661–1675*, Volume 1, ed. H. W. Turnbull (London: The Royal Society at the University Press, 1959), 416. See also Maria Popova, "Standing on the Shoulders of Giants: The Story behind Newton's Famous Metaphor of How Knowledge Progresses," Brain Pickings, February 16, 2016, https://www.brainpickings.org/2016/02/16/newton -standing-on-the-shoulders-of-giants.

29. In American academia, a faculty member's school almost always owns the rights to any "intellectual property" the faculty member invents. Patents are only granted to actual human individuals but they can be—and in most of American academia, automatically and instantly *are*—assigned to universities, institutes, corporations, and others. Thus, the UC system owns Doudna's patent rights, and the Broad Institute "owns" Zhang's. Things are sometimes done differently in Europe, however, and so Charpentier, but not any of her university employers, owns her own share of any patent rights. She has been content to let the UC system represent her interests.

30. Heidi Ledford, "Pivotal CRISPR Patent Battle Won by Broad Institute," *Nature*, September 10, 2018, https://www.nature.com/articles/ d41586-018-06656-y. Sarah Buhr, "China Sides with Emmanulle Charpentier and Jennifer Doudna in CRISPR Patent War," *TechCrunch*, June 19, 2017, https://techcrunch.com/2017/06/19/china-sides-with -emmanulle-charpentier-and-jennifer-doudna-in-crispr-patent-war. Kelly Servick, "Broad Institute Takes a Hit in European CRISPR Patent Struggle," *Science*, January 18, 2018, https://www.sciencemag.org/news/ 2018/01/broad-institute-takes-hit-european-crispr-patent-struggle.

31. Jacob S. Sherkow, "The CRISPR-Cas9 Patent Appeal: Where Do We Go From Here?," *The CRISPR Journal* October (2018), https://doi .org/10.1089/crispr.2018.0044; Jacob Sherkow, "The CRISPR Patent Decision Didn't Get the Science Right. That Doesn't Mean It Was Wrong," *STAT*, September 11, 2018.

32. Eric S. Lander, "The Heroes of CRISPR," *Cell* 164, no. 1–2 (2016): 18–28, https://doi.org/10.1016/j.cell.2015.12.041.

33. Tracy Vence, "'Heroes of CRISPR' Disputed," *The Scientist*, January 19, 2016, https://www.the-scientist.com/news-analysis/heroes-of-crispr -disputed-34188; Michael Eisen, "The Villain of CRISPR," *it is NOT junk*, 2016, http://www.michaeleisen.org/blog/?p=1825. Eisen starts his post with the arresting sentence, "There is something mesmerizing about an evil genius at the height of their craft, and Eric Lander is an evil genius at the height of his craft."

34. Sharon Begley, "Why Eric Lander Morphed from Science God to Punching Bag," *STAT*, January 25, 2016, https://www.statnews.com/ 2016/01/25/why-eric-lander-morphed. I hesitate to disagree with Begley, but I do on this one.

35. The Broad Institute, "For Journalists: Statement and Background on the CRISPR Patent Process, July 31, 2019 Statement on Motions," https:// www.broadinstitute.org/crispr/journalists-statement-and-background -crispr-patent-process.

36. For further discussion of the Nobel competition, the patent litigation, the history writing and their connections, see Henry T. Greely, "CRISPR, Patents, and Nobel Prizes," *Los Angeles Review of Books*, August 23, 2017, https://lareviewofbooks.org/article/crispr-patents-and-nobel-prizes.

Chapter 4

1. Henry T. Greely, "Human Genetic Enhancement: A Lawyer's View," *Medical Humanities Review* 17, no. 2 (2003): 42–46; review of Maxwell J. Mehlman, *Wondergenes: Genetic Enhancement and the Future of Society* (Bloomington: Indiana University Press, 2003); Henry T. Greely, "Genetic Modification," *Journal of the American Medical Association* 292 (2004): 1374–1375; review of Audrey R. Chapman and Mark S. Frankel, eds., *Designing Our Descendants: The Promises and Perils of Genetic Modifications* (Baltimore: Johns Hopkins University Press, 2003).

2. The story of recombinant DNA has been told in many places, including, in detail, in a biography of Berg. Errol C. Friedberg, *A Biography of*

Paul Berg: The Recombinant DNA Controversy Revisited (Singapore: World Scientific, 2014). A short description of the history is at https://www.sciencehistory.org/historical-profile/paul-berg.

3. Henry T. Greely, "Health Insurance, Employment Discrimination, and the Genetics Revolution," in *The Code of Codes: Scientific and Social Issues in the Human Genome Project*, ed. Daniel Kevles and Leroy Hood (Cambridge, MA: Harvard University Press, 1992).

4. This is also covered in the Friedberg biography of Berg, but for a short discussion, see https://www.sciencehistory.org/historical-profile/herbert-w-boyer-and-stanley-n-cohen.

5. Friedberg, *A Biography of Paul Berg*.

6. Sarah Zhang, "How the CRISPR Patent Dispute Became So Heated," *The Atlantic*, December 6, 2016, https://www.theatlantic.com/science/archive/2016/12/crispr-patent-in-court/509579.

7. The Asilomar story has also often been told. See, for example, Friedberg, *A Biography of Paul Berg*; Paul Berg, "Asilomar 1975: DNA Modification Secured," *Nature* 455, no. 7211 (2008): 290–291, https://doi.org/10.1038/455290a.

8. https://www.grc.org/about/history-of-grc.

9. Maxine Singer and Dieter Soll, "Guidelines for DNA Hybrid Molecules," *Science* 181, no. 4105 (1973): 1114.

10. Paul Berg et al., "Potential Biohazards of Recombinant DNA Molecules," *Science* 185, no. 4148 (July 26, 1974): 303.

11. Daniel J. Kevles, *The Baltimore Case: A Trial of Politics, Science, and Character*, rev. ed. (New York: W. W. Norton, 2000).

12. Alexander M. Capron and Renie Schapiro, "Remember Asilomar? Reexamining Science's Ethical and Social Responsibility," *Perspectives in Biology and Medicine* 44, no. 2 (Spring 2001), https://muse.jhu.edu/article/26038.

13. https://www.visitasilomar.com/discover/park-history. The Asilomar name did not come with the location but was chosen as a result

of contest. A Stanford student, Helen Salisbury, coined it by combining two Spanish words, "asilo" for refuge or asylum, and "mar" for sea.

14. The grounds contain guest rooms in a modest motel style with accommodations for conference-goers. When the rooms are not all occupied by a conference, they are available to the public at reasonable prices. I have been there with my wife for recreation and for conferences, including one to mark the 25th anniversary of the Asilomar meeting. See Henry T. Greely, "Human Genomics Research: New Challenges for Research Ethics," *Perspectives in Biology and Medicine* 44, no. 2: 221–229 (spring 2001). If you ever get the chance, go. It is gorgeous and peaceful.

15. Paul Berg, David Baltimore, Sydney Brenner, et al., "Summary Statement of the Asilomar Conference on Recombinant DNA Molecules" *Proceedings of the National Academy of Sciences* 72, no. 6 (June 1975): 1981–1984.

16. Capron and Shapiro, "Remember Asilomar." Some have speculated that the journalists were included as a consequence of the Watergate scandal of the previous year and other examples of secret decision-making. Ira Carmen, *Cloning and the Constitution: An Inquiry into Governmental Policymaking and Genetic Experimentation* (Madison: University of Wisconsin Press, 1986), 63.

17. In April 2019 the NIH announced that it was transforming the RAC into NExTRAC, the Novel and Exceptional Technology and Research Advisory Committee. This version has been stripped of any regulatory power or review authority over experimental protocols but is instead to be "the NIH Director's go-to advisory committee for advice and transparent discussions about the scientific, safety, ethical, and social issues associated with emerging biotechnologies." https://osp.od.nih.gov/2019/04/24/introducing-the-nextrac. The new committee, with a new charter and many new members, held its first meeting on December 5 and 6, 2019; its future is not yet clear.

18. Donald S. Frederickson, "Asilomar and Recombinant DNA: The End of the Beginning," in *Biomedical Politics*, ed. Kathi Hanna (Washington, DC: National Academies Press, 1991).

19. And, yes, Doudna's name can be read as "Do U DNA"! The name is a variant of an old English name, Dowdney, found mainly in Southern and Southwestern England, which in turn is supposedly a version of the Latin "Deodonatus," or "God given."

20. David Baltimore, Paul Berg, Michael Botchan, et al., "A Prudent Path Forward for Genomic Engineering and Germline Gene Modification," *Science* 348 (April 2015): 36–38.

21. Edward Lanphier, Fyodor Urnov, Sarah Ehlen Haecker, et al., "Don't Edit the Human Germ Line," *Nature* 519 (March 2015): 410–411. It is noteworthy that almost all of those authors were employees of Sangamo, a firm that was working hard to develop somatic cell gene therapies.

22. David Cyranoski and Sara Reardon, "Chinese Scientists Genetically Modify Human Embryos," *Nature*, April 22, 2015, https://www.nature.com/news/chinese-scientists-genetically-modify-human-embryos-1.17378.

23. Ralph J. Cicerone and Victor J. Dzau, "National Academy of Sciences and National Academy of Medicine Announce Initiative on Human Gene Editing," News from the National Academies, Washington, DC, 2015, https://www.nationalacademies.org/news/2015/05/national-academy-of-sciences-and-national-academy-of-medicine-announce-initiative-on-human-gene-editing.

24. Lisa M. Krieger, "National Academies Will Meet to Guide 'Gene Editing' Research," *San Jose Mercury News*, May 18, 2015, https://www.mercurynews.com/2015/05/18/national-academies-will-meet-to-guide-gene-editing-research.

25. "Advisory Group for Human Gene Editing Initiative Named," News from the National Academies, Washington, DC, June 15, 2015, https://www8.nationalacademies.org/onpinews/newsitem.aspx?RecordID=06152015.

26. "Chinese Academy of Sciences and Royal Society to Join in Convening International Summit on Human Gene Editing; Organizing Committee Named," News from the National Academies of Science, Engineering and Medicine, Washington, DC, September 14, 2015, https://www8.nationalacademies.org/onpinews/newsitem.aspx?RecordID=09142015.

27. https://www.nationalacademies.org/our-work/human-gene-editing -initiative.

28. https://www.nap.edu/catalog/21913/international-summit-on -human-gene-editing-a-global-discussion.

29. National Academies of Sciences, Engineering, and Medicine, "On Human Gene Editing: International Summit Statement," news release, December 3, 2015, https://www.nationalacademies.org/news/2015/12/ on-human-gene-editing-international-summit-statement.

30. National Academies of Sciences, Engineering, and Medicine, *Human Genome Editing: Science, Ethics, and Governance* (Washington, DC: National Academies Press, 2017), https://www.ncbi.nlm.nih.gov/books/ NBK447260.

31. National Academies of Sciences, Engineering, and Medicine, *Human Genome Editing*, 7–8.

32. National Academies of Sciences, Engineering, and Medicine, *Human Genome Editing*, 9.

33. National Academies of Sciences, Engineering, and Medicine, *Human Genome Editing*.

34. Barry Coller, "Ethics of Human Genome Editing," *Annual Review of Medicine* 70 (January 2019): 289–305. Jon Cohen, "Draw Clearer Lines around Human Gene Editing, Say Leaders of Chinese and US Science Academics," *Science,* December 13, 2018, https://www.sciencemag.org/ news/2018/12/draw-clearer-red-lines-around-human-gene-editing-say -leaders-chinese-and-us-science.

35. Nuffield Council on Bioethics, September 2016, http://nuffield bioethics.org/wp-content/uploads/Genome-editing-an-ethical -review.pdf.

36. Nuffield Council on Bioethics, July 17, 2018, https://www.nuffield bioethics.org/publications/genome-editing-and-human-reproduction. Another major national report was released in May 2019 by the German Ethics Council: Deutscher Ethikrat, *Intervening in the Human Germline: Opinion: Executive Summary and Recommendations* (translated by Aileen

Sharp), May 9, 2019, https://www.ethikrat.org/fileadmin/Publikationen/ Stellungnahmen/englisch/opinion-intervening-in-the-human-germline -summary.pdf. This has a thorough and excellent analysis, but as it came after the He experiment, I will discuss it later in this book.

37. Sharon Begley, "As a Genome Editing Summit Opens in Hong Kong, Questions Abound over China, and Why It Quietly Bowed Out," *STAT*, November 26, 2018, https://www.statnews.com/2018/11/26/human -genome-editing-summit-china.

Chapter 5

1. The declaration was first adopted in 1964 in Helsinki. The latest version is that as amended at the WMA meeting in Fortaleza, Brazil, in 2013. World Medical Association, "MA Declaration of Helsinki— Ethical Principles for Medical Research Involving Human Subjects," 2013, https://www.wma.net/policies-post/wma-declaration-of-helsinki -ethical-principles-for-medical-research-involving-human-subjects.

2. Council for International Organizations of Medical Sciences, "International Ethical Guidelines for Health-related Research Involving Humans, 4th Edition," 2016, https://cioms.ch/wp-content/uploads/2017/01/WEB -CIOMS-EthicalGuidelines.pdf.

3. UNESCO, "Universal Declaration on Bioethics and Human Rights," 2006, https://unesdoc.unesco.org/ark:/48223/pf0000146180.

4. *Trials of War Criminals before the Nuernberg Military Tribunals under Control Council Law*, Vol. II, no. 10, (October 1946–April 1949). See also The Nuremberg Trials Project, "NMT Case 1: U.S.A. v. Karl Brandt et al.: The Doctors' Trial," Harvard Law School Library, 2016.

5. *Trials of War Criminals before the Nuernberg Military Tribunals*.

6. Council for International Organizations of Medical Sciences, "International Ethical Guidelines," preamble.

7. "International Council for Harmonisation of Technical Requirements for Pharmaceuticals for Human Use," https://www.ich.org.

8. "ICH Members and Observers," ICH, https://www.ich.org/page/members-observers.

9. See, e.g., Department of Health and Human Services, 45 C.F.R § 46 (2018).

10. 21 U.S.C. § 331. Unapproved distribution of "biological products" is banned by a different statute, the Public Health Service Act of 1944 as amended, and, in any event, FDA takes the position that (just about) any biological product is also a drug.

11. 21 C.F.R § 312.

12. "FDA Regulation of Human Cells, Tissues, and Cellular and Tissue-Based Products (HCT/P's) Product List," U.S. Food and Drug Administration, https://www.fda.gov/vaccines-blood-biologics/tissue-tissue-products/fda-regulation-human-cells-tissues-and-cellular-and-tissue-based-products-hctps-product-list.

13. U.S. Food and Drug Administration, "Refuse to File: NDA and BLA Submissions to CDER, Guidance for Industry," December 2017.

14. In 2011, HHS overruled FDA for the first time ever. HHS Secretary Kathleen Sebelius disallowed emergency contraceptives from being sold over the counter to teenagers 16 or younger, even though FDA Commissioner Margaret Hamburg had issued a statement saying that it would be safe to allow this. Gardiner Harris, "Plan to Widen Availability of Morning-After Pill Is Rejected," *New York Times*, December 7, 2011, https://www.nytimes.com/2011/12/08/health/policy/sebelius-overrules-fda-on-freer-sale-of-emergency-contraceptives.html.

15. Consolidated Appropriations Act, Pub. L. No. 114–113, Stat. 2283, § 749 (2016).

16. Sara Reardon, "US Congress Moves to Block Human-Embryo Editing," *Nature*, June 25, 2015, https://www.nature.com/news/us-congress-moves-to-block-human-embryo-editing-1.17858.

17. Reardon, "U.S. Congress Moves to Block Human-Embryo Editing." Reardon quotes Professor Patti Zettler, my former student and Center for Law and the Bioscience fellow, as making the point about automatic

approval of INDs unless blocked, but I think we probably both told the reporter that independently (though my memory may be wrong, of course).

18. The Dickey-Wicker Amendment was initially included in the Balanced Budget Downpayment Act, Pub. L. No. 104–99, § 128 (1996). This section "[p]rohibits any funds made available in PL 104–91 from being used for: (1) the creation of a human embryo or embryos for research purposes; or (2) research in which a human embryo or embryos are destroyed, discarded, or knowingly subjected to risk of injury or death greater than that allowed for research under applicable Federal regulations."

19. It is interesting that the rider deals only with INDs. It does not include NDAs or BLAs. Now, without substantial clinical trial data, FDA will never approve clinical use of human germline genome editing (or anything else), but it is at least conceivable that data from clinical trials performed in other countries might eventually be submitted to FDA to justify a clinical approval. More likely, long before then the legal status of the procedure will have become clearer and more substantive, either for or against its use.

20. The story is told in Jocelyn Kaiser, "Update: House Spending Panel Restores U.S. Ban on Gene-Edited Babies," *Science* blog, https://www.sciencemag.org/news/2019/06/update-house-spending-panel-restores-us-ban-gene-edited-babies.

21. Much of the impetus came from Harvard law professor Glenn Cohen. See Emily Mullin, "Patient Advocates and Scientists Launch Push to Lift Ban on 'Three-Parent IVF,'" *STAT*, https://www.statnews.com/2019/04/16/mitochondrial-replacement-three-parent-ivf-ban. See also Eli Y. Adashi, Arthur L. Caplan, Alexander Capron, et al., "In Support of Mitochondrial Replacement Therapy, Nat. Medicine," *Nature Medicine* 25, no. 6 (June 3, 2019), https://doi.org/10.1038/s41591-019-0477-4. (I am one of the 22 mainly bioethics signatories to that letter.)

22. Human Fertilisation and Embryology Act of 1990, §§ 3ZA 2, 3, 4 (permitted eggs, sperm, and embryos, respectively).

23. Gretchen Vogel and Erik Stokstad, "U.K. Parliament Approves Controversial Three-Parent Mitochondrial Gene Therapy," *Science*, February 3, 2016, https://www.sciencemag.org/news/2015/02/uk-parliament -approves-controversial-three-parent-mitochondrial-gene-therapy.

24. Human Fertilisation and Embryology Act of 1990, § 3, Schedule 2.

25. Alice Park, "U.K. Approves First Studies of New Gene Editing Technique CRISPR on Human Embryos," February 1, 2016, https://time .com/4200695/crispr-new-gene-editing-on-human-embryos-approved. The license was renewed in 2018. Human Fertilisation and Embryology Authority, "HFEA Approves Licence Application to Use Gene Editing in Research," January 2, 2018, https://www.hfea.gov.uk/about-us/news-and -press-releases/2016-news-and-press-releases/hfea-approves-licence -application-to-use-gene-editing-in-research.

26. R. Isasi, E. Kleiderman, and B. M. Knoppers, "Editing Policy to Fit the Genome?," *Science* 351, no. 6271 (2016): 337–339, https://doi.org/ 10.1126/science.aad6778.

27. "Convention for the Protection of Human Rights and Dignity of the Human Being with regard to the Application of Biology and Medicine: Convention on Human Rights and Biomedicine, ETS No. 164," 1997.

28. Its members include the European microstates of Monaco, Andorra, San Marino, and Lichtenstein, as well as five countries on or over Europe's edge: Turkey, Georgia, Armenia, Azerbaijan, and Cyprus. Among European countries, it excludes Belarus and Kazakhstan, because of human rights concerns, and the theocratic Vatican City.

29. For the history of the convention, see Roberto Andorno, "The Oviedo Convention: A European Legal Framework at the Intersection of Human Rights and Health Law," *Bioethics* 2, no. 4 (2005): 133–143.

30. Andorno, "The Oviedo Convention."

31. "Chart of Signatures and Ratifications of Treaty 164," Council of Europe, https://www.coe.int/en/web/conventions/full-list/-/conventions/ treaty/164/signatures.

32. Interestingly, the Oviedo Convention was approved three months before and opened for signatures six weeks after the announcement of the birth of Dolly, the cloned sheep. This led to its almost immediate amendment with a new "Additional Protocol to the Convention for the Protection of Human Rights and Dignity of the Human Being with regard to the Application of Biology and Medicine, on the Prohibition of Cloning Human Beings," which opened for signatures on December 1998. This protocol states, "Any intervention seeking to create a human being genetically identical to another human being, whether living or dead, is prohibited." It has one definition: "For the purpose of this article, the term human being 'genetically identical' to another human being means a human being sharing with another the same nuclear gene set." "Additional Protocol to the Convention for the Protection of Human Rights and Dignity of the Human Being with Regard to the Application of Biology and Medicine, on the Prohibition of Cloning Human Beings, ETS No. 168," 1998.

33. Tetsuya Ishii and Yuri Hibino, "Mitochondrial Manipulation in Fertility Clinics: Regulation and Responsibility," *Reproductive Biomedicine and Society Online* 5 (2018): 93–109, https://doi.org/10.1016/j.rbms.2018.01.002, table 2.

34. Antonio Regalado, Twitter post, January 21, 2019, 5:23 a.m.

Chapter 6

1. Antonio Regalado, "Exclusive: Chinese Scientists Are Creating CRISPR Babies," *MIT Technology Review*, November 25, 2018, https://www.technologyreview.com/s/612458/exclusive-chinese-scientists-are-creating-crispr-babies.

2. Regalado, "Exclusive."

3. "Been Withdrawn with the Reason of the Original Applicants Cannot Provide the Individual Participants Data for Reviewing Safety and Validity Evaluation of HIV Immune Gene CCR5 Gene Editing in Human Embryos," Chinese Clinical Trial Registry, 2018, http://www.chictr.org.cn/showprojen.aspx?proj=32758. The listing now has this ominous title but it continues to contain some information about the trial.

4. Regalado, "Exclusive."

5. Marilynn Marchione, "Chinese Researcher Claims First Gene-Edited Babies," Associated Press, 2018, https://www.apnews.com/4997bb7 aa36c45449b488e19ac83e86d. The story was datelined Hong Kong, November 26, but 9:48 p.m., November 25, on the U.S. East Coast was 10:48 a.m. Monday, November 26, in Hong Kong.

6. Marchione, "Chinese Researcher Claims First Gene-Edited Babies."

7. Marchione, "Chinese Researcher Claims First Gene-Edited Babies."

8. Jon Cohen, "The Untold Story of the 'Circle of Trust' behind the World's First Gene-Edited Babies," *Science*, August 1, 2019, https://www .sciencemag.org/news/2019/08/untold-story-circle-trust-behind-world-s -first-gene-edited-babies. There is no indication that He was planning gene editing embryos in 2015; presumably, he hired the public relations firm for his companies or other work.

9. The He Lab, "About Lulu and Nana: Twin Girls Born Healthy after Gene Surgery as Single-Cell Embryos," 2018, https://www.youtube.com/ watch?v=th0vnOmFltc&t=7s; The He Lab, "'Designer Baby' Is an Epithet," 2018, https://www.youtube.com/watch?v=Qv1svMfaTWU&t=4s; The He Lab, "Why We Chose HIV and CCR5 First," 2018, https://www .youtube.com/watch?v=aezxaOn0efE&t=24s; The He Lab, "Draft Ethical Principles of Therapeutic Assisted Reproductive Technologies", 2018, https://www.youtube.com/watch?v=MyNHpMoPkIg&t=3s; The He Lab, "Gene Surgery in Embryos: An Embryologist Explains How It Works," 2018, https://www.youtube.com/watch?v=-1mivZUXgNI&t=7s.

10. The He Lab, "Gene Surgery in Embryos."

11. Cohen, "The Untold Story."

12. He Jiankui, Ryan Ferrell, Chen Yuanlin, et al., "Draft Ethical Principles for Therapeutic Assisted Reproductive Technologies," *The CRISPR Journal* 1, no. 6 (2018): 4–6, https://doi.org/10.1089/crispr.2018.0051. The five authors on the article are shown as affiliated with He's institution, the Southern University of Science and Technology in Shenzhen; Qin, the embryologist, also has an appointment with the Department

of Reproductive Medicine Center, Third Affiliated Hospital of Shenzhen University. The second author, Ferrell, is the American public relations consultant—see Ed Yong, "The CRISPR Baby Scandal Gets Worse by the Day," *The Atlantic*, December 3, 2018, https://www.theatlantic .com/science/archive/2018/12/15-worrying-things-about-crispr-babies -scandal/577234—and He's spokesman—see David Cyranoski, "First CRISPR Babies: Six Questions That Remain," *Nature*, November 30, 2018.

13. Email communication from Kevin Davies to Henry T. Greely (January 21, 2019).

14. Julianna LeMieux, "He Jiankui's Germline Editing Ethics Article Retracted by The CRISPR Journal," *Genetic Engineering & Biotechnology News*, February 20, 2019, https://www.genengnews.com/featured/he -jiankuis-germline-editing-ethics-article-retracted-by-the-crispr-journal.

15. Jiankui et al., "Draft Ethical Principles."

16. Chinese characters contained in the original and presumably meaning the same as the English words are omitted.

17. Cohen, "The Untold Story."

18. Pam Belluck, "Gene-Edited Babies: What a Chinese Scientist Told an American Mentor," *New York Times*, April 14, 2019, https://www .nytimes.com/2019/04/14/health/gene-editing-babies.html.

19. "Session 3—Human Embryo Editing, Second International Summit on Human Genome Editing—November 28, 2018: Day 2," National Academies of Sciences, Engineering, and Medicine, 2018, https://www .nationalacademies.org/event/11-27-2018/second-international-summit -on-human-gene-editing. On May 21, 2019, the Academies released a summary of the Second International Summit. National Academies of Sciences, Engineering, and Medicine, "Second International Summit on Human Genome Editing: Continuing the Global Discussion. Proceedings of a Workshop—in Brief," https://doi.org/10.17226/25343. (I was one of the reviewers on this short report.)

20. Robin Lovell-Badge, "CRISPR Babies: A View from the Centre of the Storm," The Company of Biologists, https://dev.biologists.org/content/ develop/146/3/dev175778.full.pdf.

21. Antonio Regalado, "Years before CRISPR Babies, This Man Was the First to Edit Human Embryos," *MIT Technology Review*, December 11, 2018, https://www.technologyreview.com/s/612554/years-before-crispr-babies -this-man-was-the-first-to-edit-human-embryos.

22. Lovell-Badge, "CRISPR Babies."

23. Pam Belluck, "How to Stop Rogue Gene-Editing of Human Embryos?," January 23, 2019, https://www.nytimes.com/2019/01/23/health/gene -editing-babies-crispr.html.

24. Belluck, "How to Stop Rogue Gene-Editing."

25. Cohen, "The Untold Story."

26. A traveler from California to Hong Kong arrives, even on a non-stop flight (of more than 15 hours), about 31 hours later than departing because of the date change at the international date line; thus, if Doudna left San Francisco on a Saturday late-night departure, she would arrive at the Hong Kong airport early on Monday morning.

27. Except for its last sentence, this paragraph is based on an email from Anne-Marie Mazza, dated June 6, 2020, on file with the author.

28. Lovell-Badge, "CRISPR Babies."

29. Cohen, "The Untold Story."

30. Sharon Begley and Andrew Joseph, "The CRISPR Shocker: How Genome-Editing Scientist He Jiankui Rose from Obscurity to Stun the World," *STAT*, December 17, 2018, https://www.statnews.com/2018/12/ 17/crispr-shocker-genome-editing-scientist-he-jiankui.

31. Email from R. Alta Charo, June 4, 2020, on file with the author.

32. Lovell-Badge, "CRISPR Babies."

33. Jon Cohen, "After Last Week's Shock, Scientists Scramble to Prevent More Gene-Edited Babies," *Science*, December 4, 2018, https://www .sciencemag.org/news/2018/12/after-last-weeks-shock-scientists-scramble -prevent-more-gene-edited-babies.

34. Cohen, "The Untold Story." Charo tells me that she had gone over each of the criteria with him, one by one, and that his answers were often confused or confusing. Charo, email.

35. The earliest tweet on the statement is time-stamped 12:56 p.m. on November 26. The poster works in Washington, DC, which leads me to conclude that the posting time shown is EST. See Greg Folkers, Twitter post, November 26, 2018, 9:56 a.m., https://twitter.com/greg_folkers/status/1067114895651557376.

36. "Statement from the Organizing Committee on Reported Human Embryo Genome Editing," Hong Kong, 2018, http://www8.nationalacademies.org/onpinews/newsitem.aspx?RecordID=11262018&_ga=2.54844584.2030215638.1543254820-484965627.1540998587.

37. Cohen, "The Untold Story."

38. The video recording of the session can be viewed at "Session 3—Human Embryo Editing, Second International Summit on Human Genome Editing—November 28, 2018: Day 2," https://www.nationalacademies.org/gene-editing/2nd_summit/second_day/index.htm.

39. Dennis Normile, "Researcher Who Created CRISPR Twins Defends His Work but Leaves Many Questions Unanswered," *Science*, November 28, 2018, https://www.sciencemag.org/news/2018/11/researcher-who-created-crispr-twins-defends-his-work-leaves-many-questions-unanswered.

40. J.K. He, "Jiankui He Talking about Human Genome Editing," 2017, https://www.youtube.com/watch?v=llxNRGMxyCc&feature=youtu.be&t=652.

Chapter 7

1. Marilynn Marchione, "Chinese Researcher Claims First Gene-Edited Babies," Associated Press, 2018, https://www.apnews.com/4997bb7aa36c45449b488e19ac83e86d.

2. "On Human Genome Editing II: Statement by the Organizing Committee of the Second International Summit on Human Genome Editing," November 29, 2018, http://www8.nationalacademies.org/

onpinews/newsitem.aspx?RecordID=11282018b&_ga=2.241822785
.21631665.1543473766-946872498.1543313092.

3. Ed Yong, "The CRISPR Baby Scandal Gets Worse by the Day," *The Atlantic*, December 3, 2018, https://www.theatlantic.com/science/archive/2018/12/15-worrying-things-about-crispr-babies-scandal/577234.

4. Christina Farr, "Experiments to Gene-Edit Babies Are 'Criminally Reckless,' Says Stanford Bio-ethicist," CNBC, November 26, 2018, https://www.cnbc.com/2018/11/26/chinese-crispr-baby-gene-editing-criminally-reckless-bio-ethicist.html.

5. Lauran Neergaard and Malcolm Ritter, "Q&A on Scientist's Bombshell Claim of Gene-Edited Babies," Associated Press, November 26, 2018, https://www.apnews.com/69c325fc818d4da0902357595a602238.

6. Jon Cohen, "'I Feel an Obligation to be Balanced:' Noted Biologist Comes to Defense of Gene Editing Babies," *Science*, November 28, 2018, https://www.sciencemag.org/news/2018/11/i-feel-obligation-be-balanced-noted-biologist-comes-defense-gene-editing-babies.

7. http://arep.med.harvard.edu/gmc/tech.html.

8. George Church and Ed Regis, *Regenesis: How Synthetic Biology Will Reinvent Nature and Ourselves* (New York: Basic Books, 2012).

9. Philip Bethge and Johan Grolle, "Interview with George Church: Can Neanderthals Be Brought Back from the Dead?," Spiegel Online, January 18, 2013, https://www.spiegel.de/international/zeitgeist/george-church-explains-how-dna-will-be-construction-material-of-the-future-a-877634.html. See also David Klinghoffer, "An Apology for Harvard's George Church (of Neanderthal Baby Fame)?," *Evolution News*, January 23, 2013, https://evolutionnews.org/2013/01/an_apology_for_.

10. Alexandra Sifferlin, "Scientists Announce Plan to Create Virus-Proof Cells," *Time*, May 1, 2018, https://time.com/5261777/scientists-virus-proof-cells.

11. Victor Dzau, Marcia McNutt, and Chunli Bai, "Wake-Up Call from Hong Kong," *Science* 362 (December 2018): 1215.

12. Jing-Bao Nie, "He Jiankui's Genetic Misadventure: Why Him? Why China?," 2018, https://www.thehastingscenter.org/jiankuis-genetic -misadventure-china. ("In the Chinese context, the declaration that his project would make China the world's first in this area is too obvious to be mentioned directly. Indeed, an initial short report of He's research appeared on the website of chief official newspaper *People's Daily* on November 26, titled 'The World's First Gene-edited Babies Genetically Resistant to AIDS Were Born in China.' It hailed He's venture as 'a milestone accomplishment *China* has achieved in the area of gene-editing technologies' (italics added). While still available on other websites, the report was soon removed, possibly due to the international as well as domestic outcry.")

13. Akshat Rathi and Echo Huang, "More Than 100 Chinese Scientists Have Condemned the CRISPR Baby Experiment as 'Crazy,'" *Quartz*, November 26, 2018, https://qz.com/1474530/chinese-scientists-condemn -crispr-baby-experiment-as-crazy. This source contains this translation of the whole letter into English.

14. Ma Danmeng, Mao Kexin, Coco Fend, and Noelle Mateer, *Baby Gene-Editing Breakthrough Claim Slammed*, *Caixin*, November 26, 2018, https://www.caixinglobal.com/2018-11-26/baby-gene-editing-break through-claim-slammed-101352172.html; Christian Shepherd and John Ruwitch, "Scientists, Officials in China Abhor Gene Editing That Geneticist Claims," Reuters, November 26, 2018, https://www.reuters.com/ article/us-health-china-babies-genes-letter/scientists-officials-in-china -abhor-gene-editing-that-geneticist-claims-idUSKCN1NW0A7; Elizabeth Cheung, "Chinese Expert in Bioethics Slams Mainland Scientist He Jiankui Who Claims to Have Created the World's First Gene-Edited Children," *South China Morning Post*, November 27, 2018, https://www .scmp.com/news/hong-kong/health-environment/article/2175273/ chinese-expert-bioethics-slams-mainland-scientist.

15. Antonio Regalado, "The Chinese Scientist Who Claims He Made CRISPR Babies Is under Investigation," *MIT Technology Review*, November 26, 2018, https://www.technologyreview.com/s/612466/the-chinese -scientist-who-claims-he-made-crispr-babies-has-been-suspended -without-pay.

16. Lily Kuo, "Work on Gene-Edited Babies Blatant Violation of the Law, Says China," *The Guardian*, November 29, 2018, https://www .theguardian.com/science/2018/nov/29/work-on-gene-edited-babies -blatant-violation-of-the-law-says-china.

17. Matthew Campbell, *China Shrinks from the Gattaca Age*, Bloomberg, December 5, 2018, https://www.bloomberg.com/news/articles/2018-12 -05/china-fiercely-decries-he-jiankui-s-human-gene-editing.

18. David Grossman, "The Infamous CRISPR Baby Scientist Is Missing," *Popular Mechanics*, December 3, 2018, https://www.popularmechanics .com/science/health/a25383837/crispr-baby-scientist-he-missing.

19. Elsie Chen and Paul Mozur, "Chinese Scientist Who Claimed to Make Genetically Edited Babies Is Kept under Guard," *New York Times*, December 28, 2018, https://www.nytimes.com/2018/12/28/world/asia/ he-jiankui-china-scientist-gene-editing.html.

20. Sharon Begley, "'CRISPR Babies' Scientist: 'I'm Actually Doing Quite Well,'" *STAT*, January 9, 2019, https://www.statnews.com/2019/01/09/ crispr-babies-scientist-im-actually-doing-quite-well.

21. Xinhua, "Guangdong Releases Preliminary Investigation Result of Gene-Edited Babies," *XINHUANET*, January 21, 2019, http://www .xinhuanet.com/english/2019-01/21/c_137762633.htm.

22. David Cyranoski, "CRISPR-Baby Scientist Fired by University," *Nature* January 22, 2019, https://www.nature.com/articles/d41586-019 -00246-2.

23. Cyranoski, "CRISPR-Baby Scientist Fired by University."

24. "China Focus: Three Jailed in China's 'Gene-Edited Babies' Trial," December 30, 2019, http://www.xinhuanet.com/english/2019-12/30/c _138667350.htm.

25. See, e.g., Antonio Regalado, "He Jiankui Faces Three Years in Prison for CRISPR Babies," *MIT Technology Review*, December 30, 2019, https://www.technologyreview.com/s/614997/he-jiankui-sentenced-to -three-years-in-prison-for-crispr-babies; Dennis Normile, "Chinese Sci- entist Who Produced Genetically-Altered Babies Sentenced to 3 Years

in Jail," *Science*, December 30, 2019, https://www.sciencemag.org/ news/2019/12/chinese-scientist-who-produced-genetically-altered -babies-sentenced-3-years-jail; Andrew Joseph, "He Jiankui, Who Created the World's First CRISPR'd Babies, Sentenced to Three Years in Prison," *STAT*, December 30, 2019, https://www.statnews.com/2019/12/30/ he-jiankui-who-created-worlds-first-crisprd-babies-sentenced-to-three -years-in-prison-for-illegal-medical-practice; Martin Farrer, "He Jiankui, Chinese Scientist Who Edited Babies' Genes, Jailed for Three Years," *The Guardian*, December 30, 2019, https://www.theguardian.com/world/ 2019/dec/30/gene-editing-chinese-scientist-he-jiankui-jailed-three -years; Sui-Lee Wee, "Chinese Scientist Who Genetically Edited Babies Gets Three Years in Prison," *New York Times*, December 30, 2019, https:// www.nytimes.com/2019/12/30/business/china-scientist-genetic-baby -prison.html.

26. Wang Pan, Xiao Sisi, and Zhou Ying, "Scientific Research Is Difficult to Hide the Fact That Illegal Medical Practice, Fame and Fortune Motivate Malicious Evasion of Supervision: Focus on the 'Gene-Edited Baby' Case," December 30, 2019, http://www.xinhuanet.com/2019-12/ 30/c_1125405443.htm.

27. CCTV, "Good Morning: He Jiankui Sentenced for Three Years in Illegal Medical Practice," December 30, 2019, http://www.court.gov.cn/ zixun-xiangqing-213381.html (in Chinese; translation is by Google). A version in English with slight substantive differences, and without this exact quotation, can be found at http://www.xinhuanet.com/ english/2019-12/30/c_138667350.htm.

28. It is not clear what saying the sentence is suspended for two years in China means. Normally, in the United States, a suspended sentence means that the convicted defendant is on probation during the period of the suspended sentence and, if his record during that time is clean, the sentence will be dismissed. I suspect that is what it means here, although between differences in criminal justice systems and uncertainties in translations, it may mean that he will start serving the 18-month sentence in two years.

29. For more information on Zhang and Qin, see Regalado, "He Jiankui Faces Three Years in Prison."

30. CCTV, "Good Morning."

31. Xinhua, "China Focus: Three Jailed in China's 'Gene-Edited Babies' Trial," Xinhuanet, December 30, 2019, http://www.xinhuanet.com/english/2019-12/30/c_138667350.htm.

32. Josiah Zayner, "CRISPR Babies Scientist He Jiankui Should Not Be Villainized," *STAT*, January 2, 2020, https://www.statnews.com/2020/01/02/crispr-babies-scientist-he-jiankui-should-not-be-villainized.

33. Dennis Normile, "Chinese Scientist Who Produced Genetically-Altered Babies Sentenced to 3 Years in Jail."

Chapter 8

1. United States Senate, "'Communists in Government Service,' McCarthy Says," February 9, 1950, https://www.senate.gov/artandhistory/history/minute/Communists_In_Government_Service.htm.

2. Rebecca Robbins, "UC Berkeley Researcher, Told of CRISPR'd Baby Study a Year Ago, Warned Scientist Not to Do It, *STAT*, November 27, 2018, https://www.statnews.com/2018/11/27/uc-berkeley-gene-edited-babies-china.

3. Candice Choi and Marilynn Marchione, "US Nobelist Was Told of Gene-Edited Babies," Associated Press, January 28, 2019, https://www.apnews.com/3f3bdc73e7c84fe685f2813510329d62.

4. "Quake Lab," Sanford University, https://quakelab.stanford.edu.

5. Andrew Joseph, Rebecca Robbins, and Sharon Begley, "An Outsider Claimed to Make Genome-Editing History—and the World Snapped to Attention," *STAT* (November 26, 2018), https://www.statnews.com/2018/11/26/he-jiankui-gene-edited-babies-china.

6. Marilynn Marchione and Christina Larson, "Could Anyone Have Stopped Gene-Edited Babies Experiment?," Associated Press, December 2, 2018, https://www.apnews.com/8d79b8da09624aabbec28d1227650a66.

7. Antonio Regalado, "Stanford Will Investigate Its Role in the Chinese CRISPR Baby Debacle," *MIT Technology Review*, February 7, 2019,

https://www.technologyreview.com/s/612892/crispr-baby-stanford-investigation.

8. Pam Belluck, "Gene-Edited Babies: What a Chinese Scientist Told an American Mentor," *New York Times*, April 14, 2019, https://www.nytimes.com/2019/04/14/health/gene-editing-babies.html.

9. His work, and Verinata, a company he founded to do this work, since acquired by Illumina, has revolutionized prenatal testing for fetuses of pregnant women by greatly lessening the need for invasive procedures like amniocentesis or chorionic villus sampling in favor of a simple blood draw. See Henry T. Greely, "Get Ready for the Flood of Fetal Gene Screening," *Nature* 469 (January 2011): 289–291. Coincidentally, the credit for inventing prenatal noninvasive testing based on fetal DNA in maternal serum is contested between Quake and Chinese researcher Dennis Lo, who was Oxford at the same time as Quake—there is a group photograph of people at Oxford that includes both of them as young students.

10. Euan A. Ashley, Matthew T. Wheeler, Atul J. Butte, et al., "Clinical Assessment Incorporating a Personal Genome," *The Lancet*, 375 (May 1, 2010): 1525–1535; and Kelly E. Ormond, Matthew T. Wheeler, Louanne Hudgins, et al., "Challenges in the Clinical Application of Whole-Genome Sequencing," *The Lancet* 375 (May 15, 2010): 1749–1751.

11. I have met Porteus perhaps 10 or 12 times. I like him but cannot say I know him well—not well enough to write a *CRISPR People* box about him.

12. "Porteus Lab, Division of Stem Cell Transplantation and Regenerative Medicine," Stanford Medicine, http://med.stanford.edu/porteuslab.html.

13. Alex Lash, "'JK Told Me He Was Planning This': A CRISPR Baby Q&A with Matt Porteus," Xconomy, December 4, 2018, https://xconomy.com/national/2018/12/04/jk-told-me-he-was-planning-this-a-crispr-baby-qa-with-matt-porteus.

14. "Editing the Genomes of Human Embryos—A Discussion with Hank Greely, William Hurlbut, and Matt Porteus," Stanford Center

for Law and the Biosciences, 2019, https://www.youtube.com/watch?v=Db6SQgsp6Zo.

15. "William Hurlbut," Wu Tsai Neurosciences Institute, Stanford University, https://neuroscience.stanford.edu/people/william-hurlbut.

16. Sharon Begley, "He Took a Crash Course in Bioethics. Then He Created CRISPR Babies," *STAT*, November 27, 2018, https://www.statnews.com/2018/11/27/crispr-babies-creator-soaked-up-bioethics.

17. Jon Cohen, "The Untold Story of the 'Circle of Trust.' behind the World's First Gene-Edited Babies," *Science*, August 1, 2019, https://www.sciencemag.org/news/2019/08/untold-story-circle-trust-behind-world-s-first-gene-edited-babies.

18. Marilynn Marchione, "One Year Later, Mystery Surrounds China's Edited Babies," Associated Press, November 26, 2019, https://apnews.com/86e4e98f7db64d0a9448e621a298ca58.

19. Marilynn Marchione, "Stanford Probes Faculty Ties to China Gene-Edited Baby Work," Associated Press, February 7, 2019, https://www.apnews.com/8480105385f64ccf98c88d1f809a8bed.

20. Regalado, "Stanford Will Investigate Its Role in the Chinese CRISPR Baby Debacle."

21. Pam Belluck, "Stanford Clears Professor of Helping with Gene-Edited Babies Experiment," *New York Times*, April 16, 2019, https://www.nytimes.com/2019/04/16/health/stanford-gene-editing-babies.html.

22. Cohen, "The Untold Story."

23. "Stanford Statement on Fact-Finding Review Related to Dr. Jiankui He," April 16, 2019, https://news.stanford.edu/2019/04/16/stanford-statement-fact-finding-review-related-dr-jiankui.

24. I am a Stanford faculty member, but I was not *that* one.

25. Henry T. Greely, "CRISPR'd Babies: Human Germline Genome Editing in the 'He Jiankui Affair,'" *Journal of Law and the Biosciences* 6, no. 1 (2019): 111–183.

26. Clive Thompson, "How to Farm Stem Cells without Losing Your Soul," *Wired*, June 1, 2005, https://www.wired.com/2005/06/stemcells.

27. "Ben Hurlbut," Arizona State University School of Life Sciences, https://sols.asu.edu/ben-hurlbut.

28. J. Benjamin Hurlbut, *Experiment in Democracy: Human Embryo Research and the Politics of Bioethics* (New York: Columbia University Press, 2017).

29. Ryan Cross, Rick Mullin , Megha Satyanarayana, et al., "As Claims of CRISPR Use in First Gene-Edited Babies Emerge, Scientists and Ethicists Respond," *Chemical & Engineering News,* November 30, 2018, https://cen.acs.org/policy/claims-CRISPR-use-first-gene/96/i48.

30. Begley, "He Took a Crash Course in Bioethics."

31. Sheila Jasanoff and J. Benjamin Hurlbut, "A Global Observatory for Gene Editing," *Nature* 555 (2018): 435–437.

32. Cohen, "The Untold Story."

33. The article notes in a correction that although Ferrell is no longer working for He, he is working for He's wife, though it doesn't say what she needs a public relations specialist for if not for her husband.

34. Ariana Eunjung Cha, "This Fertility Doctor Is Pushing the Boundaries of Human Reproduction, with Little Regulation," *Washington Post*, May 14, 2018, https://www.washingtonpost.com/national/health-science/this-fertility-doctor-is-pushing-the-boundaries-of-human-reproduction-with-little-regulation/2018/05/11/ea9105dc-1831-11e8-8b08-027a6ccb38eb_story.html.

35. Rob Stein, "Clinic Claims Success in Making Babies with 3 Parents' DNA," NPR, June 6, 2018, https://www.npr.org/sections/health-shots/2018/06/06/615909572/inside-the-ukrainian-clinic-making-3-parent-babies-for-women-who-are-infertile?t=1577914595754.

36. The article also describes Doudna as part of He's "Circle of Trust" based on her earlier interactions with him, while admitting that she only learned of He's experiment a few days before the Hong Kong Summit.

That seems to meet the article's definition of the "Circle of Trust" as people who knew or suspected what He was doing before it was publicly known by only four days and an overly strict application of the test.

37. Musunuru is also interesting because he is, as far as I know, the first author to get to publication with a book about the He Jiankui affair. Kirin Musunuru, *The CRISPR Generation: The Story of the World's First Gene-Edited Babies* (2019, self-published). I don't begrudge him his first-place finish; I do wish his book hadn't come out in the week when I was desperately trying to finish this one! The discussion of what Musunuru did and knew in the next few paragraphs is drawn from his book.

38. Musunuru, *The CRISPR Generation*.

39. Two more interesting things come out of Musunuru's discussion of the manuscript. According to Musunuru, the article said the babies were born "normal and healthy in November 2018," when we know they were born quite premature sometime in October 2018 and were hospitalized for weeks. Apparently, He was lying to a journal. Second, Musunuru never says what journal the manuscript had been submitted to. He does say, without reference (the book, which is generally good, lacks any references), it was later reported to have been *Nature* and he presumes that was the one he saw.

40. "Michael W. Deem, Ph.D.," Rice department of bioengineering, https://bioengineering.rice.edu/people/faculty/michael_deem.

41. Marilynn Marchione, "Chinese Researcher Claims First Gene-Edited Babies," Associated Press, November 26, 2018, https://www.apnews.com/4997bb7aa36c45449b488e19ac83e86d.

42. Marilynn Marchione, "Gene-Edited Baby Claim by Chinese Scientist Sparks Outrage," Associated Press, November 26, 2018, https://apnews.com/45ae0c2b32cc488fb4be717dbc71e95a.

43. I have spoken on the phone several times, occasionally at length, with Qiu, and she has quoted me before on this topic, so I may be biased in her favor. For what it's worth, I think her reporting on Deem has been exceptional.

44. Andrew Joseph, "Rice University Opens Investigation into Researcher Who Worked on CRISPR'd Baby Project," *STAT*, November 26, 2018, https://www.statnews.com/2018/11/26/rice-university-opens-investigation-into-researcher-who-worked-on-crisprd-baby-project; Todd Ackerman, "Chinese Scientist, Assisted by Rice Professor, Claims First Gene-Edited Babies," Associated Press, November 26, 2018, https://apnews.com/94fecf56cac841639d3bc9bad8db1698.

45. Ackerman, "Chinese Scientist, Assisted by Rice Professor."

46. Todd Ackerman, "Lawyers Say Rice Professor Not Involved in Controversial Gene-Edited Babies Research," *Houston Chronicle*, December 13, 2018, https://www.houstonchronicle.com/news/houston-texas/houston/article/Lawyers-say-Rice-professor-not-involved-in-13465277.php.

47. Ackerman, "Lawyers Say Rice Professor Not Involved."

48. Marchione and Larson, "Could Anyone Have Stopped Gene-Edited Babies Experiment?"

49. Sharon Begley, "Ethical Issues Plagued Newly Surfaced Paper by 'CRISPR Babies' Scientist," *STAT*, December 10, 2018, https://www.statnews.com/2018/12/10/crispr-babies-scientists-paper-rejected.

50. Jane Qiu, "American Scientist Played More Active Role in 'CRISPR Babies' Project than Previously Known," *STAT*, January 31, 2019, https://www.statnews.com/2019/01/31/crispr-babies-michael-deem-rice-he-jiankui/?_ga=2.109532678.2070114259.1551254097-2104796637.1551254097.

51. Qiu, "American Scientist Played More Active Role."

52. Antonio Regalado, "China's CRISPR Babies: Read Exclusive Excerpts from the Unseen Original Research," *MIT Technology Review*, December 2019, https://www.technologyreview.com/s/614764/chinas-crispr-babies-read-exclusive-excerpts-he-jiankui-paper.

53. Musunuru, *The CRISPR Generation*.

54. Ernest Beutler, "The Cline Affair," *Molecular Therapy* 4 (November 2001): 396–397.

55. Qiu, "American Scientist Played More Active Role."

56. Qiu, "American Scientist Played More Active Role ("Deem's possible involvement in the CRISPR babies experiment has led the Hong Kong university to review the contract, which is now 'pending on the result of the investigation undergoing at the Rice University,' said the Hong Kong university's press office.")

57. Begley, "Ethical Issues Plagued Newly Surfaced Paper."

58. Qiu, "American Scientist Played More Active Role."

59. Qiu, "American Scientist Played More Active Role."

60. Regalado, "China's CRISPR Babies: Read Exclusive Excerpts from the Unseen Original Research."

61. *The Princess Bride*, directed by Rob Reiner (20th Century Fox, 1987).

Chapter 9

1. Christina Farr, "Experiments to Gene-Edit Babies Are 'Criminally Reckless,' Says Stanford Bio-Ethicist," CNBC, November 26, 2018, https://www.cnbc.com/2018/11/26/chinese-crispr-baby-gene-editing -criminally-reckless-bio-ethicist.html; Yong, "The CRISPR Baby Scandal Gets Worse by the Day".

2. "The Nuremberg Code," United States Holocaust Memorial Museum, https://www.ushmm.org/information/exhibitions/online-exhibitions/ special-focus/doctors-trial/nuremberg-code.

3. "The Doctors Trial: The Medical Case of the Subsequent Nuremberg Proceedings, United States Holocaust Memorial Museum," United States Holocaust Memorial Museum, https://www.ushmm.org/information/ exhibitions/online-exhibitions/special-focus/doctors-trial.

4. World Medical Association, "WMA Declaration of Helsinki—Ethical Principles for Medical Research Involving Human Subjects," as amended October 2013, https://www.wma.net/policies-post/wma-declaration-of -helsinki-ethical-principles-for-medical-research-involving-human -subjects.

5. 45 CFR § 46.111(a)(2).

6. 45 CFR § 46.111(a)(1).

7. Xiao-Hui Zhang, Louis Y. Tee, Xiao-Gang Wang, et al., "Off-Target Effects in CRISPR/Cas9-Mediated Genome Engineering," *Molecular Therapy—Nucleic Acids* 4, no. 11 (2015): e264, https://doi.org/10.1038/mtna.2015.37.

8. For a discussion of this possibility, see Xiao-Jiang Li, Zhuchi Tu, Weili Yang, et al., "CRISPR: Established Editor of Human Embryos," *Cell Stem Cell* 21, no. 3 (2017): 295–296, https://doi.org/10.1016/j.stem.2017.08.007.

9. Ha Youn Shin, Chaochen Wang, Hye Kyung Lee, et al., "CRISPR/Cas9 Targeting Events Cause Complex Deletions and Insertions at 17 Sites in the Mouse Genome," *Nature Communications* 8(May 2017): 1–10, https://doi.org/10.1038/ncomms15464.

10. J.K. He, "Jiankui He Talking about Human Genome Editing," July 29, 2017, https://www.youtube.com/watch?v=llxNRGMxyCc&feature=youtu.be&t=652.

11. Benhur Lee, Matthew Sharron, Luis J. Montaner, et al., "Quantification of CD4, CCR5, and CXCR4 Levels on Lymphocyte Subsets, Dendritic Cells, and Differentially Conditioned Monocyte-Derived Macrophages," *Proceedings of the National Academy of Sciences* 96 (1999): 5215–5220; Andrew V, Albright, Joseph T. C. Shieh, Takayuki Itoh, et al., "Microglia Express CCR5, CXCR4, and CCR3, but of These, CCR5 Is the Principal Coreceptor for Human Immunodeficiency Virus Type 1 Dementia Isolates," *Journal of Virology* 73, no. 1 (1999): 205–213; Ai-Ping Jiang, Jin-Feng Jiang, Ming-Gao Guo, et al., "Human Blood-Circulating Basophils Capture HIV-1 and Mediate Viral Trans-Infection of CD4+ T Cells," *Journal of Virology* 89, no. 15 (2015): 8050–8062.

12. Gail G. Vaday, Pournima A. Kadam, Donna M. Peehl, et al., "Expression of CCL5 (RANTES) and CCR5 in Prostate Cancer," *Prostate* 66, no. 2 (2006): 124–134, https://doi.org/10.1002/pros.20306; Santos Mañes, Emilia Mira, Ramón Colomer, et al., "CCR5 Expression Influences the Progression of Human Breast Cancer in a P53-Dependent Manner,"

Journal of Experimental Medicine 198, no. 9 (2003): 1381–1389, https://doi.org/10.1084/jem.20030580; Daniela Sicoli, Xuanmao Jiao, Xiaoming Ju, et al., "CCR5 Receptor Antagonists Block Metastasis to Bone of V-Src Oncogene–Transformed Metastatic Prostate Cancer Cell Lines," *Cancer Research* 73 (2014): 7103–7114.

13. *CCR5* or its close relative, *CCR2*, was found in 10 out of 11 species (wallabies were the only exception), from lizards to humans, though not in any fish. Hisayuki Nomiyama, Naoki Osada, and Osamu Yoshie, "Systematic Classification of Vertebrate Chemokines Based on Conserved Synteny and Evolutionary History," *Genes to Cells* 18 (2013): 1–16, https://doi.org/10.1111/gtc.12013.

14. Ute V. Solloch, Kathrin Lang, Vinzenz Lange, et al., "Frequencies of Gene Variant CCR5-Δ32 in 87 Countries Based on Next-Generation Sequencing of 1.3 Million Individuals Sampled from 3 National DKMS Donor Centers," *Human Immunology* 78 (2017): 710–717, https://doi.org/10.1016/j.humimm.2017.10.001.

15. William G. Glass, David H. McDermott, Jean K. Lim, et al., "CCR5 Deficiency Increases Risk of Symptomatic West Nile Virus Infection," *Journal of Experimental Medicine* 203 (2006): 35–40, https://doi.org/10.1084/jem.20051970.

16. A. Falcon, M. T. Cuevas, A. Rodriguez-Frandsen, et al., "CCR5 Deficiency Predisposes to Fatal Outcome in Influenza Virus Infection," *Journal of General Virology* 96, no. 8 (August 1, 2015): 2074–2078, https://doi.org/10.1099/vir.0.000165.

17. "Flu Kills 646,000 People Worldwide Each Year: Study," MedicineNet, https://www.medicinenet.com/script/main/art.asp?articlekey=208914; A. Danielle Iuliano, Katherine M. Roguski, Howard H. Chang, et al., "Estimates of Global Seasonal Influenza-Associated Respiratory Mortality: A Modelling Study," *The Lancet* 391 (2018): 1285–1300, https://doi.org/10.1016/S0140-6736(17)33293-2; "Number of Deaths Due to HIV/AIDS," World Health Organization, https://www.who.int/gho/hiv/epidemic_status/deaths_text/en; "Global Statistics," Hiv.gov, https://www.hiv.gov/hiv-basics/overview/data-and-trends/global-statistics.

18. Xinchun Yu, Chunfang Wang, Tao Chen, et al., "Excess Pneumonia and Influenza Mortality Attributable to Seasonal Influenza in Subtropical Shanghai, China," *BMC Infectious Diseases* 17 (2017): 756, https://doi.org/10.1186/s12879-017-2863-1.

19. National Health and Family Planning Commission of the People's Republic of China, "2015 China AIDS Response Progress Report," 2015, 9, https://www.unaids.org/sites/default/files/country/documents/CHN _narrative_report_2015.pdf.

20. In early June 2019 it looked like we did have some evidence and that there were substantial negative effects. *Nature Medicine* published a study of over 400,000 people in the U.K. Biobank that seemed to show that people with two copies of the *CCR5Δ32* variation had a 21 percent increase in mortality compared to people with no such variation. Xinzhu Wei and Rasmus Nielsen, "CCR5-Δ32 Is Deleterious in the Homozygous State in Humans," *Nature Medicine*, 2019, https://doi.org/10.1038/s41591 -019-0459-6. Since then, however, another group, headed by David Reich, has shown that the first article was based on a technical problem with their analysis and that a proper calculation showed no such effect. Robert Maier, Ali Akbari, Xinzhu Wei, et al., "No Statistical Evidence for an Effect of CCR5-Δ32 on Lifespan in the UK Biobank Cohort," *BioRxiv*, January 1, 2019, https://doi.org/10.1101/787986. The initial authors then retracted it. Xinzhu Wei and Rasmus Nielsen, "Retraction Note: CCR5- Δ32 Is Deleterious in the Homozygous State in Humans," *Nature Medicine* 25, no. 11 (2019): 1796, https://doi.org/10.1038/s41591-019-0637-6.

21. John Novembre, Alison P Galvani, and Montgomery Slatkin, "The Geographic Spread of the CCR5 Δ32 HIV-Resistance Allele," *PLOS Biology* 3, no. 11 (October 18, 2005): e339, https://doi.org/10.1371/journal .pbio.0030339.

22. Maryam Zafer, Hacsi Horvath, Okeoma Mmeje, et al., "Effectiveness of Semen Washing to Prevent HIV Transmission and Assist Pregnancy in HIV-Discordant Couples: A Systematic Review and Meta-Analysis," *Fertility and Sterility* 105, no. 3 (2016): 645–655, https://doi.org/10.1016/j .fertnstert.2015.11.028.

23. Cédric Blanpain, Frédérick Libert, Gilbert Vassart, et al., "CCR5 and HIV Infection," *Receptors and Channels* 8 (2002): 19–31, https://doi.org/10.1080/10606820212135.

24. "HIV and AIDS in China," Avert, October 3, 2019, https://www.avert.org/professionals/hiv-around-world/asia-pacific/china; "Country Comparison—HIV/AIDS Prevalence Rate," The World Factbook, https://www.cia.gov/library/publications/the-world-factbook/rankorder/2155rank.html. The infection rate in China is on the rise, however. Sifan Zheng, "The Growing Threat of China's HIV Epidemic," *The Lancet* 3, no. 7 (2018), https://doi.org/https://doi.org/10.1016/S2468-2667(18)30098-7; "Reported Cases of HIV in China Are Rising Rapidly," *The Economist*, January 12, 2019, https://www.economist.com/china/2019/01/12/reported-cases-of-hiv-in-china-are-rising-rapidly.

25. Henry T. Greely, "He Jiankui, Embryo Editing, CCR5, the London Patient, and Jumping to Conclusions," *STAT*, April 15, 2015, https://www.statnews.com/2019/04/15/jiankui-embryo-editing-ccr5.

26. Jinzhou Qin, Yangran Chen, Shuo Song, et al., "Birth of Twins after Genome Editing for HIV Resistance" (unpublished, 2018). See also Antonio Regalado, "China's CRISPR Babies: Read Exclusive Excerpts from the Unseen Original Research," *MIT Technology Review*, December 2019, https://www.technologyreview.com/s/614764/chinas-crispr-babies-read-exclusive-excerpts-he-jiankui-paper, noting He and his coauthors described the mutation as conferring resistance to HIV-1 similar to "the natural CCR5 Δ32 variation."

27. Afam A. Okoye and Louis J. Picker, "CD4(+) T-Cell Depletion in HIV Infection: Mechanisms of Immunological Failure," *Immunological Review* 254 (2013): 54–64, https://doi.org/10.1111/imr.12066.

28. Tsutomu Murakami and Naoki Yamamoto, "Role of CXCR4 in HIV Infection and Its Potential as a Therapeutic Target," *Future Microbiology* 5 (2010): 1025–1039, https://doi.org/10.2217/fmb.10.67.

29. Olivier Manches, Davor Frleta, and Nina Bhardwaj, "Dendritic Cells in Progression and Pathology of HIV Infection," *Trends in Immunology* 35 (2014): 114–122, https://doi.org/10.1016/j.it.2013.10.003.

30. Laura Waters, Sundhiya Mandalia, Paul Randell, et al., "The Impact of HIV Tropism on Decreases in CD4 Cell Count, Clinical Progression, and Subsequent Response to a First Antiretroviral Therapy Regimen," *Clinical Infectious Diseases* 46 (2008): 1617–1623, https://doi.org/10.1086/587660.

31. Ariel D. Weinberger and Alan S. Perelson, "Persistence and Emergence of X4 Virus in HIV Infection," *Mathematical Biosciences and Engineering* 8, no. 2 (2011): 605–626, https://doi.org/10.3934/mbe.2011.8.605; "X4-Tropic Virus," U.S. Department of Health and Human Services AIDSinfo, https://aidsinfo.nih.gov/understanding-hiv-aids/glossary/882/x4-tropic-virus.

32. Shi-hua Xiang, Beatriz Pacheco, Dane Bowder, et al., "No Characterization of a Dual-Tropic Human Immunodeficiency Virus (HIV-1) Strain Derived from the Prototypical X4 Isolate HXBc2," *Virology* 438, no. 1 (2013): 5–13, https://doi.org/10.1016/j.virol.2013.01.002.

33. Jacqueline D. Reeves, Sam Hibbitts, Graham Simmons, et al., "Primary Human Immunodeficiency Virus Type 2 (HIV-2) Isolates Infect CD4-Negative Cells via CCR5 and CXCR4: Comparison with HIV-1 and Simian Immunodeficiency Virus and Relevance to Cell Tropism in Vivo," *Journal of Virology* 73, no. 9 (1999): 7795–7804, https://www.ncbi.nlm.nih.gov/pmc/articles/PMC104307.

34. Mary Engel, "Timothy Ray Brown: The Accidental AIDS Icon," Fred Hutch, February 20, 2015, https://www.fredhutch.org/en/news/center-news/2015/02/aids-icon-timothy-ray-brown.html; Gero Hütter, Daniel Nowak, Maximilian Mossner, et al., "Long-Term Control of HIV by CCR5 Delta32/Delta32 Stem-Cell Transplantation," *The New England Journal of Medicine* 360 (2009): 692–698, https://doi.org/10.1056/NEJMoa0802905; Jon Cohen, "Has a Second Person with HIV Been Cured?," *Science*, March 4, 2019, https://www.sciencemag.org/news/2019/03/has-second-person-hiv-been-cured.

35. Ravindra K. Gupta, Sultan Abdul-Jawad, Laura E. McCoy, et al., "HIV-1 Remission Following CCR5Δ32/Δ32 Haematopoietic Stem-Cell Transplantation," *Nature* 568, no. 7751 (2019): 244–248, https://doi.org/

10.1038/s41586-019-1027-4. See the discussion in Cohen, "Has a Second Person with HIV Been Cured?"

36. Carolyn Y. Johnson, "A Decade after the First Person Was Cured of HIV, a Second Patient Is in Long-Term Remission," *Washington Post*, March 5, 2019, https://www.washingtonpost.com/health/2019/03/05/decade-after-first-person-was-cured-hiv-second-patient-is-long-term-remission.

37. Apoorva Mandavilli, "H.I.V. Is Reported Cured in a Second Patient, a Milestone in the Global AIDS Epidemic," *New York Times*, March 4, 2019, https://www.nytimes.com/2019/03/04/health/aids-cure-london-patient.html.

38. Gupta et al., "HIV-1 Remission Following CCR5Δ32/Δ32 Haematopoietic Stem-Cell Transplantation."

39. Lambros Kordelas, Jens Verheyen, Stefan Esser, et al., "Shift of HIV Tropism in Stem-Cell Transplantation with CCR5 Delta32 Mutation," *New England Journal of Medicine* 371, no. 9 (2014): 880–882, https://doi.org/10.1056/NEJMc1405805; Melissa Healy, "Two Patients with HIV Are in Remission. How Many More Will Follow Them?," *Los Angeles Times*, March 6, 2019, https://www.latimes.com/science/sciencenow/la-sci-sn-hiv-patient-remission-future-20190306-story.html.

40. Ed Yong, "A Reckless and Needless Use of Human Gene Editing on Human Embryos," *The Atlantic*, November 26, 2018, https://www.theatlantic.com/science/archive/2018/11/first-gene-edited-babies-have-allegedly-been-born-in-china/576661; Yong, "The CRISPR Baby Scandal Gets Worse by the Day."

41. David Cyranowski and Heidi Ledford, "Genome-Edited Baby Claim Provokes International Outcry," *Nature*, November 26, 2018, https://www.nature.com/articles/d41586-018-07545-0.

42. John Lauerman and Naomi Kresge, "Gene-Edited Twins in China Still Face Risk of HIV Infection," Bloomberg, November 27, 2018, https://www.bloomberg.com/news/articles/2018-11-27/gene-edited-twins-in-china-still-face-risk-of-hiv-infection.

43. Musunuru, Kiran, Twitter post. November 26, 2019, 2:33 a.m., https://twitter.com/kiranmusunuru/status/1199275014039703552.

44. Antonio Regalado, "China's CRISPR Twins Might Have Had Their Brains Inadvertently Enhanced," *MIT Technology Review*, February 21, 2019, https://www.technologyreview.com/s/612997/the-crispr-twins-had-their-brains-altered.

45. Miou Zhou, Stuart Greenhill, Shan Huang, et al., "CCR5 Is a Suppressor for Cortical Plasticity and Hippocampal Learning and Memory," *ELife* 5 (December 2016): 1–30, https://doi.org/10.7554/eLife.20985.

46. Mary T. Joy, Einor Ben Assayag, Dalia Shabashov-Stone, et al., "CCR5 Is a Therapeutic Target for Recovery after Stroke and Traumatic Brain Injury," *Cell* 176, no. 5 (2019): 1143–1157.e13, https://doi.org/10.1016/j.cell.2019.01.044. Regalado's article notes that the study, which looked at patients in Tel Aviv, had some data showing the people with one *CCR5Δ32* variant had obtained more education than those with two normal copies. This was not a finding of the paper and, in light of other significant differences between the two groups—notably that nearly 90 percent of the patients with the rare variant were of Ashkenazic ancestry compared with only 57 percent of those without that gene version—that evidence seems very susceptible to cultural or other interpretations.

47. Regalado, "China's CRISPR Twins Might Have Had Their Brains Inadvertently Enhanced."

48. 45 C.F.R. 46.201 et seq. "Fetus" is defined in 45 C.F.R. § 46.202(c).

49. 45 C.F.R. Subpart D, §§46.101 et seq.

50. 21 C.F.R. § 56.111(a)(2).

51. World Medical Association, "WMA Declaration of Helsinki—Ethical Principles for Medical Research Involving Human Subjects," Principle 17.

52. George J. Annas, "How Did Claims of CRISPR Babies Hijack an International Gene-Editing Summit?," December 4, 2018, https://www

.bu.edu/sph/2018/12/04/how-did-claims-of-crispr-babies-hijack-an
-international-gene-editing-summit.

53. The quotation is from the form itself. This can no longer be reached at the Chinese site where it was first posted, that for the Southern University of Science and Technology, He's (former) employer, but an English translation can be found at "Informed Consent," https://www.sciencemag .org/sites/default/files/crispr_informed-consent.pdf. See also Derek Lowe, "After Such Knowledge," In the Pipeline, November 26, 2018, https://blogs .sciencemag.org/pipeline/archives/2018/11/28/after-such-knowledge.

54. These refer to the failure to become pregnant or sustain a pregnancy after two efforts or the aftermath of a pregnancy where the embryo or fetus has "genetic defects or other serious disease."

55. "Session 3—Human Embryo Editing, Second International Summit on Human Genome Editing—November 28, 2018: Day 2," National Academies of Sciences, Engineering, and Medicine, 2018, https://www .nationalacademies.org/event/11-27-2018/second-international-summit -on-human-gene-editing.

56. Marilynn Marchione, "Gene-Edited Baby Claim by Chinese Scientist Sparks Outrage," Associated Press, November 26, 2018.

57. "Session 3—Human Embryo Editing, Second International Summit on Human Genome Editing—November 28, 2018: Day 2."

58. Christopher VanLang, "Should the Twin Babies That Chinese Scientist He Jiankui Gene-Edited Be Able to Sue Him, and Possibly Their Parents as Well, for Altering Their DNA? (Answered)," Quora, December 3, 2018, https://www.quora.com/Should-the-twin-babies-that-Chinese -scientist-He-Jiankui-gene-edited-be-able-to-sue-him-and-possibly-their -parents-as-well-for-altering-their-DNA.

59. "Session 3—Human Embryo Editing, Second International Summit on Human Genome Editing—November 28, 2018: Day 2."

60. Xinhua, "Guangdong Releases Preliminary Investigation Result of Gene-Edited Babies," Xinhuanet, January 21, 2019, http://www.xinhuanet .com/english/2019-01/21/c_137762633.htm.

61. Suzanne Statline and Ian Sample, "Scientist in China Defends Human Embryo Gene Editing," *The Guardian*, November 28, 2018, https://www.theguardian.com/science/2018/nov/28/scientist-in-china-defends-human-embryo-gene-editing.

62. Ezekiel Emanuel, "Ending Concerns about Undue Inducement," *Journal of Law, Medicine, & Ethics* 32 (2004): 100–105, https://doi.org/10.1111/j.1748-720X.2004.tb00453.x.

63. Southern University of Science and Technology Statement on the Genetic Editing of Human Embryos Conducted by Dr. Jiankui He," http://sustc.edu.cn/en/info_focus/2871.

64. "Session 3—Human Embryo Editing, Second International Summit on Human Genome Editing—November 28, 2018: Day 2.

65. Rita Liao, "Hospital in China Denies Links to World's First Gene-Edited Babies," TechCrunch, November 26, 2018, https://techcrunch.com/2018/11/26/hospital-denies-gene-edited-babies-china.

66. Marilynn Marchione, "Chinese Researcher Claims First Gene-Edited Babies," Associated Press, November 26, 2018, https://www.apnews.com/4997bb7aa36c45449b488e19ac83e86d.

67. Yong, "The CRISPR Baby Scandal Gets Worse by the Day."

68. Harmonicare Medical Holdings Ltd, "Clarification Announcement Regarding Certain Media Reports," November 27, 2018, https://www.sciencemag.org/sites/default/files/crispr_clarification.pdf.

69. Xinhua, "Guangdong Releases Preliminary Investigation Result of Gene-Edited Babies."

70. Xinhua, "Guangdong Releases Preliminary Investigation Result of Gene-Edited Babies."

71. Henry T. Greely, "CRISPR'd Babies: Human Germline Genome Editing in the 'He Jiankui Affair,'" *Journal of Law and the Biosciences* 6, no. 1 (2019): 111–183; Marilynn Marchione, "Stanford Probes Faculty Ties to China Gene-Edited Baby Work," Associated Press, February 7, 2019, https://www.apnews.com/8480105385f64ccf98c88d1f809a8bed.

72. "Declaration of Independence: A Transcription," https://www.archives.gov/founding-docs/declaration-transcript.

73. National Academies of Sciences, Engineering, and Medicine et al., *Human Genome Editing: Science, Ethics, and Governance*, Washington, DC, 2017.

74. Nuffield Council on Bioethics, *Genome Editing and Human Reproduction*, 2018.

75. Elizabeth Cooney, "What We Know—and What We Don't—about the Claim of the World's First Gene-Edited Babies," *STAT*, November 26, 2018, https://www.statnews.com/2018/11/26/what-we-know-gene-edited-babies-crispr.

76. Paul Simon, "The Boxer" (1969).

Chapter 10

1. The discussion about how Science needs to ostracize He, encouraging snitching, and express humility grew out of my work in Henry T. Greely, "How Should Science Respond to CRISP'd Babies?," *Issues in Science and Technology* 35, no. 3 (Spring 2019) (2019): 32–37.

2. Ernest Beutler, "The Cline Affair," *Molecular Therapy* 4 (2001): 396–397.

3. Choe Sang-Hun, "Korean Scientist's New Project: Rebuild after Cloning Disgrace," *New York Times*, 2014, https://www.nytimes.com/2014/03/01/world/asia/scientists-new-project-rebuild-after-cloning-disgrace.html.

4. I am not the only one to make such a suggestion. In December 2019, Soren H. Hough and Ayokunmi Ajetunmobi, writing in a letter to the editor of *The CRISPR Journal*, called for a "boycott of academics who would perform or advise on clinical germline-editing experiments." Soren H. Hough and Ayokunmi Ajetunmobi, "A CRISPR Moratorium Isn't Enough: We Need a Boycott," *The CRISPR Journal* 2, no. 6, published online on December 16, 2019, https://www.liebertpub.com/doi/full/10.1089/crispr.2019.0041.

5. Sam Kean, "The Soviet Era's Deadliest Scientist Is Regaining Popularity in Russia," December 19, 2017, https://www.theatlantic.com/science/archive/2017/12/trofim-lysenko-soviet-union-russia/548786.

6. This is the translation of Voltaire's line in his novel *Candide* on the execution of British admiral John Byng after losing the battle of Minorca. In Portsmouth, Candide witnesses the execution of an officer by firing squad and is told that "in this country, it is good to kill an admiral from time to time, in order to encourage the others." Voltaire, and Daniel Gordon, *Candide* (Boston: Bedford/St. Martin's, 1999); "John Byng," https://en.wikipedia.org/wiki/John_Byng.

7. Victor J. Dzau, Marcia McNutt, and Chunli Bai, "Wake-up Call from Hong Kong," *Science* 362, no. 6420 (2018): 1215, https://doi.org/10.1126/science.aaw3127.

8. David Freeman Engstrom, "Private Enforcement's Pathways: Lessons from Qui Tam Litigation," *Columbia Law Review* 114 (2014): 1913–2006.

9. E.g., see United States v. Johnson, 546 F.2d 1225, 1227 (5th Cir. 1977); Lancey v. United States, 356 F.2d 407, 410 (9th Cir. 1966).

10. Elder Consumer Protection Program, "Mandatory Reporting Statues for Elder Abuse," Center for Excellence in Elder Law, Stetson Law, 2016, https://www.stetson.edu/law/academics/elder/home/media/Mandatory-reporting-Statutes-for-elder-abuse-2016.pdf. In California, e.g., "Any person who has assumed full or intermittent responsibility for the care or custody of an elder or dependent adult whether or not he or she receives compensation" must report "physical abuse . . . , abandonment, abduction, isolation, financial abuse, or neglect" to a range of government offices "immediately or as soon as practically possible." Failure to report is a misdemeanor, punishable by not more than six months in the county jail, or by a fine of not more than $1,000, or by both imprisonment and fine. Elder Abuse and Dependent Adult Civil Protection Act, California Welfare & Institutions Code § 15630.

11. The Federal Child Abuse Prevention and Treatment Act "requires each State to have provisions or procedures for requiring certain individuals to report known or suspected instances of child abuse and

neglect." 42 U.S.C. § 5106a(b)(2)(B)(i). Additionally, states have a variety of statutes. Child Welfare Information Gateway, "Mandatory Reporters of Child Abuse and Neglect," 2019, https://www.childwelfare.gov/topics/systemwide/laws-policies/statutes/manda. In California, e.g., a law makes over 50 classes of professionals mandatory reporters, including health care workers, educators, clergy, and others. Shouse California Law Group, "California's Mandatory Reproting Law of Child Abuse and Child Neglect," https://www.shouselaw.com/mandatory-reporting#2.

The penalties for nonreporting are normally six months in jail, a $1,000 fine, or both, though they increase if the child died or was subject to great bodily harm. The Child Abuse and Neglect Reporting Act, California Penal Code §§ 11164–11174.3.

12. Stanford Law School, "CRISPR'd Babies—A Discussion with Matt Porteus," February 7, 2019, https://www.youtube.com/watch?v=Db6S Qgsp6Zo.

13. David Baltimore, Paul Berg, Michael Botchan, et al., "A Prudent Path Forward for Genomic Engineering and Germline Gene Modification," *Science* 348, no. 6230 (2015): 36–38.

14. Organizing Committee for the International Summit on Human Gene Editing, "On Human Gene Editing: International Summit Statement," National Academies of Sciences, Engineering, and Medicine, December 5, 2015, http://www8.nationalacademies.org/onpinews/newsitem.aspx ?RecordID=12032015a.

15. National Academies of Sciences, Engineering, and Medicine et al., *Human Genome Editing: Science, Ethics, and Governance*, Washington, DC, 2017.

16. Nuffield Council on Bioethics, *Genome Editing and Human Reproduction*, 2018.

17. "Session 3—Human Embryo Editing, Second International Summit on Human Genome Editing—November 28, 2018: Day 2," National Academies of Sciences, Engineering, and Medicine, https://www .nationalacademies.org/event/11-27-2018/second-international-summit -on-human-gene-editing.

18. "On Human Genome Editing II: Statement by the Organizing Committee of the Second International Summit on Human Genome Editing," National Academies of Sciences, Engineering, and Medicine, November 29, 2018, http://www8.nationalacademies.org/onpinews/ newsitem.aspx?RecordID=11282018b&_ga=2.241822785.21631665 .1543473766-946872498.1543313092.

19. https://livestream.com/accounts/7036396/events/8464254/videos/ 184096639/player?width=640&height=360&enableInfo=true&default Drawer=&autoPlay=false&mute=false&t=1543422491964 at 26:14 to 26:40.

20. Jon Cohen, "After Last Week's Shock, Scientists Scramble to Prevent More Gene-Edited Babies," *Science*, December 4, 2018, https://www .sciencemag.org/news/2018/12/after-last-weeks-shock-scientists-scramble -prevent-more-gene-edited-babies.

21. Mary Louise Kelly, "Harvard Medical School Dean Weighs in on Ethics of Gene Editing," NPR, November 29, 2018, https://www.npr .org/2018/11/29/671996695/harvard-medical-school-dean-weighs-in-on -ethics-of-gene-editing.

22. Broad Communications, "Broad Scientists and Geneticists Discuss Issues Raised by Clinical Germline Genome Editing," November 26, 2018, https://www.broadinstitute.org/news/broad-scientists-and-geneticists -discuss-issues-raised-clinical-germline-genome-editing.

23. Robin Lovell-Badge, "CRISPR Babies: A View from the Centre of the Storm," *Development* 146, no. 3 (2019), https://doi.org/10.1242/dev .175778.

24. Victor J. Dzau, Marcia McNutt, and Chunli Bai, "Wake-up Call from Hong Kong," *Science* 362, no. 6420 (2018): 1215, https://doi.org/ 10.1126/science.aaw3127.

25. National Academies of Sciences, Engineering, and Medicine, "New International Commission Launched on Clinical Use of Heritable Human Genome Editing," May 22, 2019, http://www8.nationalacademies.org/ onpinews/newsitem.aspx?RecordID=5222019.

26. "International Commission on the Clinical Use of Human Germline Genome Editing," https://www.nationalacademies.org/gene-editing/international-commission.

27. World Health Organization, "Human Genome Editing," December 14, 2018, https://www.who.int/ethics/topics/human-genome-editing/en.

28. World Health Organization, "WHO Expert Advisory Committee on Developing Global Standards for Governance and Oversight of Human Genome Editing," https://www.who.int/ethics/topics/human-genome-editing/committee-members/en; GenomeWeb, "WHO Panel Announced," February 13, 2019, https://www.genomeweb.com/scan/who-panel-announced#.XIXc4FNKgWo.

29. Eric Lander, Françoise Baylis, Feng Zhang, et al., "Adopt a Moratorium on Heritable Genome Editing," *Nature* 567, no. 7747 (2019): 165–168, https://doi.org/10.1038/d41586-019-00726-5.

30. Dzau, McNutt, and Ramarkrishnan, "Academies Action on Germline Editing."

31. Carrie D. Wolinetz and Francis Collins, "NIH Pro Germline Editing Moratorium," *Nature* 576 (2019): 175.

32. Lexico, powered by Oxford, https://www.lexico.com/definition/moratorium.

33. Lander et al., "Adopt a Moratorium on Heritable Genome Editing."

34. Lander et al., "Adopt a Moratorium on Heritable Genome Editing."

35. Advisory Committee on Developing Global Standards for Governance and Oversight of Human Genome Editing, Report of the First Meeting (May 19, 2019), https://www.who.int/ethics/topics/human-genome-editing/GenomeEditing-FirstMeetingReport-FINAL.pdf?ua=1; see also Jon Cohen, "WHO Panel Proposes New Global Registry for all CRISPR Human Experiments," *Science* (March 29, 2019), https://www.sciencemag.org/news/2019/03/who-panel-proposes-new-global-registry-all-crispr-human-experiments.

36. Sara Reardon, "World Health Organization Panel Weighs in on CRISPR-Babies Debate," *Nature* 567 (2019): 444–445, https://www.nature.com/articles/d41586-019-00942-z.

37. Jon Cohen, "WHO Panel Proposes New Global Registry for all CRISPR Human Experiments," *Science,* March 29, 2019, https://www.sciencemag.org/news/2019/03/who-panel-proposes-new-global-registry-all-crispr-human-experiments; Advisory Committee on Developing Global Standards for Governance and Oversight of Human Genome Editing, "Report of the First Meeting," 2019.

38. Advisory Committee on Developing Global Standards, "Report."

39. "WHO-RUSH Human Genome Editing 1st Advisory Committee VPC," WHO, March 19, 2019, https://www.who.int/ethics/topics/human-genome-editing/Human-genome-editing-1st-advisory-committee-VPC.pdf?ua=1.

40. "WHO Launches Global Registry on Human Genome Editing," WHO, August 29, 2019, https://www.who.int/news-room/detail/29-08-2019-who-launches-global-registry-on-human-genome-editing.

41. David Cyranoski, "Russian Biologist Plans More CRISPR-Edited Babies," *Nature* 570 (June 10, 2019): 145–146, https://www.nature.com/articles/d41586-019-01770-x.

42. "Denis V. Rebrikov," Bulletin of RSMU, https://vestnikrgmu.ru/general_editor/?lang=en.

43. Cyranoski, "Russian Biologist Plans More CRISPR-Edited Babies."

44. Michael Le Page, "Exclusive: Five Couples Lined Up for CRISPR Babies to Avoid Deafness," *New Scientist*, July 4, 2019, https://www.newscientist.com/article/2208777-exclusive-five-couples-lined-up-for-crispr-babies-to-avoid-deafness.

45. Interestingly, mutations in this gene are associated with six different genetic conditions, not all of which involve deafness. NIH, "GJB2 Gene," Genetics Home References, https://ghr.nlm.nih.gov/gene/GJB2#conditions.

46. Le Page, "Exclusive: Five Couples Lined Up for CRISPR Babies."

47. Antonio Regalado, "Putin Could Decide for the World on CRISPR Babies," *MIT Technology Review*, September 20, 2019, https://www .technologyreview.com/f/614450/putin-could-decide-for-the-world-on -crispr-babies.

48. Olga Dobrovidova, "Calling Embryo Editing 'Premature,' Russian Authorities Seek to Ease Fears of a Scientist Going Rogue," *STAT*, October 16, 2019, https://www.statnews.com/2019/10/16/russia-health -ministry-calls-human-embryo-editing-premature.

49. David Cyranoski, "Russian 'CRISPR-Baby' Scientist Has Started Editing Genes in Human Eggs with Goal of Altering Deaf Gene, *Nature* 574 (October 24, 2019): 465–466, https://www.nature.com/articles/ d41586-019-03018-0.

50. Elena G. Grebenshchikova, "Russia's Stance on Human Genome Editing," *Nature* 575 (November 28, 2019): 596, https://www.nature.com/ articles/d41586-019-03617-x.

51. "Human Genome Editing: As We Explore Options for Global Governance, Caution Must Be Our Watchword," WHO, https://www.who.int/ ethics/topics/human-genome-editing/ethics-explore-options-for-global -governance.pdf?ua=1.

52. Meanwhile, there is some reason to think that another Chinese researcher, Hui Yang, has been talking about using genome editing to make babies. Paul Knoepfler, "Heads Up on Hui Yang, Another Potential Aspiring CRISPR Baby Researcher," The Niche Blog (January 22, 2020), https://ipscell.com/2020/01/heads-up-on-hui-yang-another-potential -aspiring-crispr-baby-researcher. On the other hand, in February 2019 Antonio Regalado published a piece about another person who was planning to do germline gene editing, not on embryos but through injecting CRISPR into men's testicles. Antonio Regalado, "The DIY Designer Baby Project Funded with Bitcoin," *MIT Technology Review*, February 1, 2019, https://www.technologyreview.com/s/612838/the-transhumanist-diy -designer-baby-funded-with-bitcoin. A quick review of the article shows that this effort should not engender serious concern.

53. Ed Yong, "Chinese Project Probes the Genetics of Genius," *Nature* (May 14, 2013), https://www.nature.com/news/chinese-project-probes-the-genetics-of-genius-1.12985.

54. David Cyranoski, "Gene-Edited 'Micropigs' to Be Sold as Pets at Chinese Institute," *Nature* 526 (September 29, 2015): 18, https://www.nature.com/news/gene-edited-micropigs-to-be-sold-as-pets-at-chinese-institute-1.18448.

55. Sarah Zhang, "Would You Buy a Genetically-Engineered Cashmere Sweater?," *The Atlantic,* October 26, 2016, https://www.theatlantic.com/health/archive/2016/10/cashmere-goat-crispr/505163.

56. Antonio Regalado, "First Gene-Edited Dogs Reported in China," *MIT Technology Review,* October 19, 2015, https://www.technologyreview.com/s/542616/first-gene-edited-dogs-reported-in-china.

57. Helen Shen, "First Monkeys with Customized Mutations Born," *Nature* (January 30, 2014), https://www.nature.com/news/first-monkeys-with-customized-mutations-born-1.14611.

58. Megan Molteni, "With Embryo Base Editing, China Gets Another CRISPR First," *Wired,* August 21, 2018, https://www.wired.com/story/crispr-base-editing-first-china.

59. Mark E. Steiner, "Inclusion and Exclusion in American Legal History," *Asian American Law Journal* 23, no. 1 (2016): 69–98.

60. Leigh Bristol-Kagan, "Chinese Migration to California, 1851–1882: Selected Industries of Work, the Chinese Institutions and the Legislative Exclusion of a Temporary Work Force." Ph.D. diss., Harvard University, Cambridge, MA, 1982.

61. Frankie Huang, "Letter: How China's Penchant for Eugenics Led to CRISPR Babies," *Caixin,* December 17, 2018, https://www.caixinglobal.com/2018-12-17/letter-how-chinas-penchant-for-eugenics-led-to-crispr-babies-101360013.html.

62. Sui-Lee Wei, "China Halts Work by Scientist Who Said He Edited Babies' Genes," *New York Times,* November 29, 2018, https://www

.nytimes.com/2018/11/29/science/gene-editing-babies-china.html; "Research Activities of Persons Halted over Gene-Edited Babies Incident," Xinhua, November 29, 2018, http://www.xinhuanet.com/english/2018 -11/29/c_137640174.htm.

63. Yanan Wang and Fu Ting, "China Drafts Rules on Biotech after Gene-Editing Scandal," Associated Press, February 27, 2019, https:// www.apnews.com/47aa8ffa382c4ae19eb6ec202f93ddf8.

64. Antonio Regalado, Twitter post, January 21, 2019, 5:32 a.m. https:// twitter.com/antonioregalado/status/1087342258721034241; Serenitie Wang, "Chinese Authorities Say World's First Gene-Edited Babies Were Illegal," CNN, January 22, 2019, https://www.cnn.com/2019/01/21/ health/china-gene-editing-babies-intl/index.html.

65. Tom Hancock and Wang Xueqiao, "China Set to Tighten Regulations on Gene-Editing Research," *Financial Times*, January 25, 2019, https://www.ft.com/content/a464bd9c-f869-11e8-af46-2022a0b02a6c.

66. Wang and Ting, "China Drafts Rules on Biotech."

67. Jane Qiu, "China Creating National Medical Ethics Committee to Oversee High-Risk Clinical Trials," *STAT*, March 5, 2019, https://www .statnews.com/2019/03/05/china-creating-national-medical-ethics-committee.

68. David Cyranoski, "China Set to Introduce Gene-Editing Regulation Following CRISPR-Baby Furore," *Nature* (May 20, 2019), https://www .nature.com/articles/d41586-019-01580-1.

69. Reed Smith, "The Adoption of the Chinese Civil Code and its Implications on Contracts Relating to China," (June 3, 2020), https:// www.reedsmith.com/en/perspectives/2020/06/the-adoption-of-the -chinese-civil-code-and-its-implications-on-contracts.

70. Ruipeng Lei, Xiaomei Zhai, Wei Zhu, et al., "Reboot Ethics Governance in China," *Nature* 569 (May 8, 2019): 184–186, https://www .nature.com/articles/d41586-019-01408-y.

71. Jon Cohen, "Chinese Bioethicists Call for 'Reboot' of Biomedical Regulation after Country's Gene-Edited Baby Scandal," *Science*, May 8, 2019,

https://www.sciencemag.org/news/2019/05/chinese-bioethicists-call
-reboot-biomedical-regulation-after-country-s-gene-edited-baby.

72. Lei et al., "Reboot Ethics Governance in China."

Chapter 11

1. Henry T. Greely, "Human Genetic Enhancement: A Lawyer's View," *Medical Humanities Review* 17, no. 2 (n.d.): 42–46; Henry T. Greely, "Genetic Modification. Book Review of Designing Our Descendants: The Promises and Perils of Genetic Modifications," *Journal of the American Medical Association* 292 (2004): 1374–1375.

2. This chapter looks at the three most prominent statements. Carolyn Brokowski provides a broader survey of more than 50 ethics statements, apparently comprehensive through the time of its 2018 publication. Carolyn Brokowski, "Do CRISPR Germline Ethics Statements Cut It?," *The CRISPR Journal* 1, no. 2 (2018): 115–125, https://doi.org/10.1089/crispr.2017.0024.

3. National Academies of Sciences, Engineering, and Medicine et al., *Human Genome Editing: Science, Ethics, and Governance*, Washington, DC, 2017.

4. Nuffield Council on Bioethics, *Genome Editing and Human Reproduction*, 2018.

5. Deutscher Ethikrat, *Intervening in the Human Germline: Opinion: Executive Summary and Recommendations*, 2019.

6. Summary of Principles and Recommendations in National Academies of Sciences, Engineering, and Medicine et al., *Human Genome Editing: Science, Ethics, and Governance*.

7. National Academies of Sciences, Engineering, and Medicine et al., 112.

8. Heritable Genome Editing in National Academies of Sciences, Engineering, and Medicine et al.

9. Heritable Genome Editing in National Academies of Sciences, Engineering, and Medicine et al.

10. Nuffield Council on Bioethics, *Genome Editing: An Ethical Review*, 2016, http://nuffieldbioethics.org/wp-content/uploads/Genome-editing -an-ethical-review.pdf.

11. Deutscher Ethikrat, *Eingriffe in die menschliche Keimbahn*, 2019.

12. Deutscher Ethikrat, *Intervening in the Human Germline: Executive Summary and Recommendations*.

13. Sharon Begley, "German Ethics Council Expresses Openness to Eventual Embryo Editing," *STAT*, May 13, 2019, https://www.statnews .com/2019/05/13/germline-editing-german-ethics-council.

14. Deutscher Ethikrat, *Intervening in the Human Germline: Executive Summary and Recommendations*, 4.

15. Deutscher Ethikrat, 20–21.

16. UNESCO, "Universal Declaration on the Human Genome and Human Rights," 1997, 3.

17. Tim Lewens, "Blurring the Germline: Genome Editing and Transgenerational Epigenetic Inheritance," *Bioethics* 34, no. 1 (2019), https:// doi.org/10.1111/bioe.12606.

18. Mareike C. Janiak, "Of Starch and Spit," *eLife* 8 (2019): e47523, https://doi.org/10.7554/eLife.47523.

19. And, in fact, there has been such a substantial increase, timed for about when substantial numbers of children with type 1 diabetes first saved by insulin had their own children old enough to develop the condition. The reasons for that increase continue, however, to be debated, as the following article shows. Edwin A.M. Gale, "The Rise of Childhood Type 1 Diabetes in the 20th Century," *Diabetes* 51, no. 12 (2002): 3353– 3361, https://doi.org/10.2337/diabetes.51.12.3353.

20. Laura Clark, "World Population by Percentage of Blood Types," World Atlas, 2019, https://www.worldatlas.com/articles/what-are-the -different-blood-types.html.

Chapter 12

1. I probably should have discussed the "unnaturalness" critique in the preceding chapter, as, if accepted, it too would make human germline genome editing inherently bad. But the last chapter focused on the interesting and somewhat novel issue of "sacralizing" the genome, treating it as a sacred. The unnaturalness contention is quite different and quite usual—it has been used to criticize innovations at least since railroads and anesthesia (especially for childbirth) and undoubtedly long before. I did not want to sully the previous chapter's argument with this, but it seems to me to fit better as one of a set of arguments, albeit as the weakest one.

2. Exodus 23:19; Exodus 34:26; Deuteronomy 14:21.

3. Daniel Kevles, *In the Name of Eugenics: Genetics and the Uses of Human Heredity* (New York: Alfred A. Knopf, 1985).

4. I have written some on these issues. See Henry T. Greely, "Remarks on Human Biological Enhancement," *University of Kansas Law Review* 56 (2008): 1139–1157; Henry T. Greely, "Regulating Human Biological Enhancements: Questionable Justifications and International Complications," *The Mind, The Body and The Law: University of Technology Sydney Law Review* 7 (2005): 87–110/*Santa Clara Journal of International Law* 4 (2006): 87–110 (joint issue); Henry T. Greely, "Disabilities, Enhancements, and the Meanings of Sport," *Stanford Law and Policy Review* 15 (2004): 99–132; Henry T. Greely, "Of Nails and Hammers: Human Biological Enhancement and American Policy Tools," in *Enhancing Human Capacities*, ed. Julian Savulescu, Ruud ter Muelen, and Guy Kahane (Hoboken, NJ: Wiley-Blackwell, 2011).

Chapter 13

1. "Genetics Home Reference: What Are the Different Ways in Which a Genetic Condition Can Be Inherited?," NIH U.S. National Library of Medicine, https://ghr.nlm.nih.gov/primer/inheritance/inheritancepatterns. Another way is called codominant: the gene for blood type O is

recessive; the genes for blood types A and B are "codominant": someone with an A and an O is type A, someone with a B and an O is type B, someone with an A and a B is AB. This inheritance pattern is not common enough to worry about.

2. Mitochondria have long been believed to be passed on to the next generation only in the mother's egg, not in the father's sperm. Recently this has been questioned. Shiyu Luo, C. Alexander Valencia, Jinglan Zhang, et al., "Biparental Inheritance of Mitochondrial DNA in Humans," *Proceedings of the National Academy of Sciences* 115, no. 51 (December 18, 2018): 13039–13044, https://doi.org/10.1073/PNAS.1810946115. Three papers have disagreed: John Vissing, "Paternal Comeback in Mitochondrial DNA Inheritance.," *Proceedings of the National Academy of Sciences of the United States of America* 116, no. 5 (January 29, 2019): 1475–1476, https://doi.org/10.1073/pnas.1821192116; Sabine Lutz-Bonengel and Walther Parson, "No Further Evidence for Paternal Leakage of Mitochondrial DNA in Humans Yet," *Proceedings of the National Academy of Sciences of the United States of America* 116, no. 6 (February 5, 2019): 1821–1822, https://doi.org/10.1073/pnas.1820533116; Antonio Salas, Sebastian Schönherr, Hans-Jürgen Bandelt, et al., "Extraordinary Claims Require Extraordinary Evidence in the Case of Asserted mtDNA Biparental Inheritance," *BioRxiv* (January 1, 2019), https://doi.org/10.1101/585752. Another has disagreed with those three and with the original claim: Sofia Annis, Zoe Fleischmann, Mark Khrapko, et al., "Quasi-Mendelian Paternal Inheritance of Mitochondrial DNA: A Notorious Artifact, or Anticipated Behavior?," *Proceedings of the National Academy of Sciences of the United States of America* 116, no. 30 (July 23, 2019): 14797–14798, https://doi.org/10.1073/pnas.1821436116. I don't know who is right (though I share the skepticism expressed in one of the titles—"Extraordinary claims require extraordinary evidence"), and it is not important for our purposes. But it does provide an example of how some apparently simple human genetics questions remain without certain answers.

3. A. H. Handyside, E. H. Kontogianni, K. Hardy, et al., "Pregnancies from Biopsied Human Preimplantation Embryos Sexed by Y-Specific DNA

Amplification," *Nature* 344 (1990): 768–770, https://doi.org/10.1038/344768a0.

4. U.S. Centers for Disease Control and Prevention, "2016 Assisted Reproductive Technology Fertility Clinic Success Rates Report," 2016, https://www.cdc.gov/art/reports/2016/fertility-clinic.html.

5. U.S. Centers for Disease Control and Prevention, "2015 Assisted Reproductive Technology Fertility Clinic Success Rates Report," 2015, https://www.cdc.gov/art/reports/archive.html.

6. Julie Steffann, Pierre Jouannet, Jean-Paul Bonnefont, et al., "Could Failure in Preimplantation Genetic Diagnosis Justify Editing the Human Embryo Genome?," *Cell Stem Cell* 22, no. 4 (2018): 481–482, https://doi.org/10.1016/j.stem.2018.01.004.

7. Henry T. Greely, *The End of Sex and the Future of Human Reproduction* (Cambridge, MA: Harvard University Press, 2016).

8. Hong Ma, Nuria Marti-Gutierrez, Sang-Wook Park, et al., "Correction of a Pathogenic Gene Mutation in Human Embryos," *Nature* 548, no. 7668 (2017): 413–419, https://doi.org/10.1038/nature23305. See also Nerges Winblad and Fredrik Lanner, "At the Heart of Gene Edits in Human Embryos," *Nature* 548, no. 7668 (2017): 398–400, https://doi.org/10.1038/nature23533.

9. Paul Knoepfler, "Counterpoints to Lovell-Badge & Daley's CRISPR Baby Rationales," Knoepfler Lab Stem Cell Blog, 2019, https://ipscell.com/2019/02/counterpoints-to-lovell-badge-daleys-crispr-baby-rationales. And note that this would not affect the embryos of parents with two copies of a genetic variation that causes an autosomal dominant condition. Their embryos would have one bad copy from them but (almost certainly) one good copy from their mates.

10. Jacob S. Sherkow, Patricia J. Zettler, and H. T. Greely, "Is It 'Gene Therapy'?, *Journal of Law & Biosciences* 5, no. 3 (2018): 786–793. https://doi.org/10.1093/jlb/lsy020.

11. "FDA Approves Novel Gene Therapy to Treat Patients with a Rare Form of Inherited Vision Loss," U.S. Food and Drug Administration, 2017,

https://www.fda.gov/news-events/press-announcements/fda-approves
-novel-gene-therapy-treat-patients-rare-form-inherited-vision-loss.

12. "Statement from FDA Commissioner Scott Gottlieb, M.D., and Peter Marks, M.D., Ph.D., Director of the Center for Biologics Evaluation and Research on New Policies to Advance Development of Safe and Effective Cell and Gene Therapies," U.S. Food and Drug Administration, 2019, https://www.fda.gov/news-events/press-announcements/statement-fda -commissioner-scott-gottlieb-md-and-peter-marks-md-phd-director -center-biologics.

13. Daniel J. Weiner, Emilie M. Wigdor, Stephan Ripke, et al., "Polygenic Transmission Disequilibrium Confirms That Common and Rare Variation Act Additively to Create Risk for Autism Spectrum Disorders," *Nature Genetics* 49, no. 7 (2017): 978–985, https://doi.org/10.1038/ng .3863.

14. "Polygenic Risk Scores," National Human Genome Research Institute, https://www.genome.gov/Health/Genomics-and-Medicine/Polygenic -risk-scores.

15. Ute V. Solloch, Kathrin Lang, Vinzenz Lange, et al., "Frequencies of Gene Variant CCR5-Δ32 in 87 Countries Based on Next-Generation Sequencing of 1.3 Million Individuals Sampled from 3 National DKMS Donor Centers," *Human Immunology* 78, no. 11–12 (2017): 710–717, https://doi.org/10.1016/j.humimm.2017.10.001.

16. Stephen S. Hall, "Genetics: A Gene of Rare Effect," *Nature* (April 9, 2013), https://www.nature.com/news/genetics-a-gene-of-rare-effect -1.12773.

17. Rahul Chaudhary, Jalaj Garg, Neeraj Shah, et al., "PCSK9 Inhibitors: A New Era of Lipid Lowering Therapy," *World Journal of Cardiology* 9, no. 2 (2017): 76, https://doi.org/10.4330/wjc.v9.i2.76.

18. The initial paper, Xinzhu Wei and Rasmus Nielsen, "CCR5-Δ32 Is Deleterious in the Homozygous State in Humans," *Nature Medicine* 25 (2019): 909–910, https://doi.org/10.1038/s41591-019-0459-6, was retracted. The paper pointing out the error, Robert Maier, Ali Akbari, Xinzhu Wei, et al., "No Statistical Evidence for an Effect of CCR5-Δ32 on

Lifespan in the UK Biobank Cohort," *BioRxiv* (January 1, 2019), https:// doi.org/10.1101/787986. And the retraction, Xinzhu. Wei and Rasmus Nielsen, "Retraction Note: CCR5-Δ32 Is Deleterious in the Homozygous State in Humans," *Nature Medicine* 25 (2019): 1796, https://doi .org/10.1038/s41591-019-0637-6.

19. A. Falcon, M. T. Cuevas, A. Rodriguez-Frandsen, et al., "CCR5 Deficiency Predisposes to Fatal Outcome in Influenza Virus Infection," *Journal of General Virology* 96, no. 8 (August 1, 2015): 2074–2078, https:// doi.org/10.1099/vir.0.000165.

20. Jean K. Lim and Philip M. Murphy, "Chemokine Control of West Nile Virus Infection," *Experimental Cell Research* 317, no. 5 (March 10, 2011): 569–574, https://doi.org/10.1016/J.YEXCR.2011.01.009.

21. Carolyn Y. Johnson, "Long Awaited Cystic Fibrosis Drug Could Turn Deadly Disease into a Manageable Condition," *The Washington Post*, October 31, 2019, https://www.washingtonpost.com/health/ 2019/10/31/long-awaited-cystic-fibrosis-drug-could-turn-deadly-disease -into-manageable-condition.

22. Henry T. Greely, "Remarks on Human Biological Enhancement," *University of Kansas Law Review* 56 (2008): 1139–1157.

23. Some might argue for mutations in the myostatin gene, which provide much heavier musculature in dogs and, in at least one case, a human, are such an exception. Dana S. Mosher, Pascale Quignon, Carlos D. Bustamante, et al., "A Mutation in the Myostatin Gene Increases Muscle Mass and Enhances Racing Performance in Heterozygote Dogs," *PLoS Genetics* 3, no. 5 (2007): 779–786, https://doi.org/10.1371/journal .pgen.0030079 (in dogs); E. Paul Zehr, "The Man of Steel, Myostatin, and Super Strength," *Scientific American*, June 14, 2013, https://blogs .scientificamerican.com/guest-blog/the-man-of-steel-myostatin-and -super-strength (humans). I'm not convinced. We don't know the long-term effects of the enhanced muscle, not just on overall health but even on effective strength. If your muscles were to become too strong for your tendons, ligaments, joints, and bones, is that enhancing?

24. "Genetics Home Reference: MC1R Gene," NIH U.S. National Library of Medicine, https://ghr.nlm.nih.gov/gene/MC1R#conditions.

25. "Doping. Bigger, Stronger, Faster: The Unexpected Virtue of Danger," Sports Ethics, 2015, https://ethicsofsports.wordpress.com/tag/doping.

26. David Epstein, *The Sports Gene* (New York: Current, 2013).

27. Tatum S. Simonson, Yingzhong Yang, Chad D. Huff, et al., "Genetic Evidence for High-Altitude Adaptation in Tibet," *Science* 329, no. 5987 (July 2, 2010): 72–75, https://doi.org/10.1126/science.1189406.

28. Haroon Siddique, "Blood Doping: What Is It and Has Anyone Died as a Result of It?," *The Guardian*, August 2, 2015, https://www.theguardian .com/sport/2015/aug/02/blood-doping-what-is-it-and-has-anyone-died -as-a-result-of-it.

29. Yuval N. Harari, *Homo Deus: A Short History of Tomorrow* (New York: Harper, 2017).

30. H. G. Wells, *The Time Machine: An Invention* (London: W. Heineman, 1895).

31. Marc Ereshefsky, "Species," in *Stanford Encyclopedia of Philosophy Archive*, ed. Edward N. Zalta, Fall 2017, https://plato.stanford.edu/ archives/fall2017/entries/species; R. L. Sandler, *The Ethics of Species: An Introduction* (Cambridge: Cambridge University Press, 2012).

32. Lee Silver, *Remaking Eden: Cloning and Beyond in a Brave New World* (New York: Avon Books, 1997).

33. Victor Frankenstein did, at least in Mary Shelley's book. (His motives in the various plays, movies, comic books, and retellings of her story in other media are usually unclear.) At one point he looked forward to a new species blessing him as their creator. Mary Shelley, *Frankenstein, or the Modern Prometheus* (London: Lackington, Hughes, Harding, Mavor & Jones, 1818). And at this point I cannot help but refer you to one of my favorites among my recent articles and not just because it is such an unusual publishing venue for me: Henry T. Greely, "Frankenstein and Modern Bioscience: Which Story Should We Heed," *Huntington Library Quarterly* 83 (2020).

Chapter 14

1. Leslie G. Biesecker and Nancy B. Spinner, "A Genomic View of Mosaicism and Human Disease," *Nature Reviews Genetics* 14, no. 5 (2013): 307–320, https://doi.org/10.1038/nrg3424.

2. National Conference of State Legislatures, "Embryonic and Fetal Research Laws," 2016, http://www.ncsl.org/issues-research/health/embryonic-and-fetal-research-laws.aspx.

3. The German Embryonenschutzgesutz, or Embryo Protection Act, criminalizes the production of embryos for any purpose other than a pregnancy. France also prohibits the "artificial production of embryos." See DRZE, "II. Selected National and International Laws and Regulations—Stem Cell Research Dossier," 2004, 1–10, http://www.drze.de/in-focus/stem-cell-research/laws-and-regulations.

4. The Dickey-Wicker Amendment has been attached to appropriations bills for the Departments of Health and Human Services, Labor, and Education since 1996. The 2019 (and latest version) states: "(a) None of the funds made available in this Act may be used for—(1) the creation of a human embryo or embryos for research purposes; or (2) research in which a human embryo or embryos are destroyed, discarded, or knowingly subjected to risk of injury or death greater than that allowed for research on fetuses in utero under 45 CFR 46.208(a)(2) and Section 498(b) of the Public Health Service Act [1](42 U.S.C. 289g(b)) (Title 42, Section 289g(b), United States Code). (b) For purposes of this section, the term "human embryo or embryos" includes any organism, not protected as a human subject under 45 CFR 46 (the Human Subject Protection regulations) . . . that is derived by fertilization, parthenogenesis, cloning, any other means from one or more human gametes (sperm or egg) or human diploid cells (cells that have two sets of chromosomes, such as somatic cells)."

5. "Egg donation is barred in China, Germany, Italy, and Norway; [while] paying women to donate eggs is prohibited in most of Europe, as well as in Canada and other nations." Paris Martineau, "Inside the Quietly Lucrative Business of Donating Human Eggs," *Wired*, April 23,

2019, https://www.wired.com/story/inside-lucrative-business-donating -human-eggs.

6. California Assembly, Reproductive health and research: oocyte procurement AB 922, 2019, https://leginfo.legislature.ca.gov/faces/bill NavClient.xhtml?bill_id=201920200AB922.

7. California Health and Safety Code §§125330–125335SB; 1260 Reproductive Health and Research (2006). http://leginfo.legislature.ca.gov/ faces/billTextClient.xhtml?bill_id=200520060SB1260. In 2013, Bill 926—a bill that would have allowed payment for the "time, discomfort, and inconvenience" of egg donation for research—passed in the California State Assembly, but was vetoed by Governor Jerry Brown. Charlotte Schubert, "California Set to Lift Restrictions on Egg Donation," *Scientific American*, June 19, 2013, https://www.scientificamerican.com/article/ california-set-to-lift-restrictions-on-egg-donation.

8. Mitinori Saitou and Hidetaka Miyauchi, "Gametogenesis from Pluripotent Stem Cells," *Cell Stem Cell* 18, no. 6 (2016): 721–735, https:// doi.org/10.1016/j.stem.2016.05.001; Chika Yamashiro, Kotaro Sasaki, Yukihiro Yabuta, et al., "Generation of Human Oogonia from Induced Pluripotent Stem Cells in Vitro," *Science* 362, no. 6412 (2018): 356–360, https://doi.org/10.1126/science.aat1674. See also I. Glenn Cohen and Alex Pearlman, "Creating Eggs and Sperm from Stem Cells: The Next Big Thing in Assisted Reproduction?," *STAT*, June 5, 2019, https://www .statnews.com/2019/06/05/creating-eggs-sperm-stem-cells.

9. Henry T. Greely, *The End of Sex and the Future of Human Reproduction* (Cambridge, MA: Harvard University Press, 2016).

10. Masahito Tachibana, Paula Amato, Michelle Sparman, et al., "Human Embryonic Stem Cells Derived by Somatic Cell Nuclear Transfer," *Cell* 153, no. 6 (2013): 1228–1238, https://doi.org/10.1016/j.cell.2013 .05.006.

11. Alessia Deglincerti, Gist F. Croft, Lauren N. Pietila, et al., "Self-Organization of the in Vitro Attached Human Embryo," *Nature* 533, no. 7602 (2016): 251–254, https://doi.org/10.1038/nature17948; Marta N. Shahbazi, Agnieszka Jedrusik, Sanna Vuoristo, et al., "Self-Organization

of the Human Embryo in the Absence of Maternal Tissues," *Nature Cell Biology* 18, no. 6 (2016): 700–708, https://doi.org/10.1038/ncb3347; Janet Rossant, "Implantation Barrier Overcome," *Nature* 533 (2016): 182–183, https://doi.org/10.1038/nature17894.

12. Deglincerti et al., "Self-Organization of the in Vitro Attached Human Embryo"; Shahbazi et al., "Self-Organization of the Human Embryo."

13. Insoo Hyun, Amy Wilkerson, and Josephine Johnston, "Embryology Policy: Revisit the 14-Day Rule," *Nature* 533, no. 7602 (2016): 169–171, https://doi.org/10.1038/533169a.

14. Robert L. Perlman, "Mouse Models of Human Disease: An Evolutionary Perspective," *Evolution, Medicine, and Public Health* 2016, no. 1 (January 2016), https://doi.org/10.1093/emph/eow014.

15. David Grimm, "Record Number of Monkeys Being Used in U.S. Research," *Science*, 2018, 1–8, https://doi.org/10.1126/science.aav9290.

16. National Institutes of Health, "Statement by NIH Director Dr. Francis Collins on the Institute of Medicine Report Addressing the Scientific Need for the Use of Chimpanzees in Research," 2019, https://www.nih.gov/news-events/news-releases/statement-nih-director-dr-francis-collins-institute-medicine-report-addressing-scientific-need-use-chimpanzees-research. See also David Grimm, "Has U.S. Biomedical Research on Chimpanzees Come to an End?," *Science*, 2015, 1–7, https://doi.org/10.1126/science.aad1635.

17. Nita A. Farahany, Henry T. Greely, Steven Hyman, et al., "The Ethics of Experimenting with Human Brain Tissue," *Nature* 556 (2018): 429–432, https://www.nature.com/articles/d41586-018-04813-x; Henry T. Greely, "The Dilemma of Human Brain Surrogates—Scientific Opportunities, Ethical Concerns," in *Neuroscience and Law: Complicated Crossings and New Perspectives*, ed. Antonio D'Aloia Errigo and Maria Chiara (New York: Springer, 2020); Henry T. Greely, "Human Brain Surrogates Research: The Onrushing Ethical Dilemma," *American Journal of Bioethics* (in press).

18. Jeff Sebo, "Should Chimpanzees Be Considered 'Persons'?," *New York Times*, April 7, 2018, https://www.nytimes.com/2018/04/07/opinion/sunday/chimps-legal-personhood.html. In a concurrence, at least one

judge from the NY Court of Appeals (the NY Supreme Court) signaled chimpanzees should possibly have certain rights typically reserved for humans. Matter of Nonhuman Rights Project, Inc. v Lavery, 31 N.Y. 3d 1054 (2018) (J. Fahey concurring).

19. Kenneth G. Gould, "Ovum Recovery and in Vitro Fertilization in the Chimpanzee," *Fertility and Sterility* 40, no. 3 (1983): 378–383, https://doi.org/10.1016/S0015-0282(16)47304-8.

20. B. D. Bavister, D. E. Boatman, K. Collins, et al., "Birth of Rhesus Monkey Infant after in Vitro Fertilization and Nonsurgical Embryo Transfer," *Proceedings of the National Academy of Sciences* 81, no. 7 (1984): 2218–2222, https://doi.org/10.1073/pnas.81.7.2218.

21. Terry Devitt, "First Test-Tube Rhesus Monkey Healthy and Virile after 15 Years," *University of Wisconsin-Madison News*, August 20, 1998, https://news.wisc.edu/first-test-tube-rhesus-monkey-healthy-and-virile-after-15-years/

22. Y. Kang, C. Chu, F. Wang, et al., "CRISPR/Cas9-mediated Genome Editing in Nonhuman Primates," *Disease Models & Mechanisms* 12, no. 10 (2019), https://dmm.biologists.org/content/12/10/dmm039982.

23. Pub. L. No. 100-578, 102 Stat. 2903 (codified as amended at 42 U.S.C. § 263a (2012)).

24. American Society for Reproductive Medicine, "Executive Summary: Oversight of Assisted Reproductive Technology," 2010, 1–12, http://www.asrm.org/globalassets/asrm/asrm-content/about-us/pdfs/oversiteofart.pdf.

Chapter 15

1. *Catechism of the Catholic Church, U.S. Catholic Church* (Washington, DC: USCCB Publishing, 1995). Section 2366 of the Catechism of the Catholic Church stresses "the inseparable connection, established by God, which man on his own initiative may not break, between the unitive significance and the procreative significance which are both inherent to the marriage act." Section 2377 condemns artificial insemination

and IVF (even without donor gametes) because "they dissociate the sexual act from the procreative act."

2. Human Fertilisation and Embryology Authority, "Code of Practice," 9th ed., 2019, https://www.hfea.gov.uk/media/2793/2019-01-03-code-of -practice-9th-edition-v2.pdf.

3. "Risk Evaluation and Mitigation Strategies," U.S. Food and Drug Administration, https://www.fda.gov/drugs/drug-safety-and-availability/risk -evaluation-and-mitigation-strategies-rems. For more general information on REMS, see the FDA's FAQs on them: "Frequently Asked Questions (FAQs) about REMS," U.S. Food and Drug Administration, https:// www.fda.gov/drugs/risk-evaluation-and-mitigation-strategies-rems/ frequently-asked-questions-faqs-about-rems.

4. I. Glenn Cohen, *Patients with Passports: Medical Tourism, Law and Ethics* (Oxford: Oxford University Press, 2015); Henry T. Greely, *The End of Sex and the Future of Human Reproduction* (Cambridge, MA: Harvard University Press, 2016), 224, 269, 283, 294–296, 303, 311–312.

5. "Citizen's Guide to U.S. Federal Law on the Extraterritorial Sexual Exploitation of Children," The U.S. Department of Justice, https://www .justice.gov/criminal-ceos/citizens-guide-us-federal-law-extraterritorial -sexual-exploitation-children.

6. European Parliament and the Council of 9 March 2011, "Directive 2011/24/EU on the Application of Patients' Rights in Cross-Border Healthcare," *Official Journal of the European Union*, 2011, https://eur-lex .europa.eu/legal-content/EN/TXT/HTML/?uri=CELEX:32011L0024&from =EN.

7. Marcia C. Inhorn, Daphna Birenbaum-Carmeli, Soraya Tremaynec, et al., "Assisted Reproduction and Middle East Kinship: A Regional and Religious Comparison," *Reproductive Biomedicine & Society Online* 4 (2017): 41–51, https://doi.org/10.1016/j.rbms.2017.06.003.

8. Draft legislation altering these rules was recently passed by the National Assembly. Sarah Gregory, "France's Lower House Approves IVF Draft Law," *BioNews*, September 30, 2019, https://www.bionews.org.uk/ page_145256.

9. Irene Riezzo, Margherita Neri, Stefania Bello, et al., "Italian Law on Medically Assisted Reproduction: Do Women's Autonomy and Health Matter?," *BMC Women's Health* 16 (2016): 44, https://doi.org/10.1186/s12905-016-0324-4.

10. The most plausible exception is when the editing is only needed on the paternal contribution and science has advanced enough to allow us to make sperm from cell lines. In that case, one might do genome editing on the sperm and then use it in artificial insemination, not IVF. Note, though, that some of the laws apply to assisted reproduction generally, not just IVF but also artificial insemination.

11. Judith Daar, *The New Eugenics: Selective Breeding in an Era of Reproductive Technologies* (New Haven, CT: Yale University Press, 2017).

12. Katie Falloon and Philip M. Rosoff, "Who Pays? Mandated Insurance Coverage for Assisted Reproductive Technology," *AMA Journal of Ethics* 16, no. 1 (2014): 63–69, https://doi.org/10.1001/virtualmentor.2014.16.1.msoc1-1401; "Medicaid's Role for Women," Kaiser Family Foundation, March 28, 2019, https://www.kff.org/womens-health-policy/fact-sheet/medicaids-role-for-women.

13. Rachel Gurevich, "How Much Does IVF Really Cost?," Verywell Family, March 20, 2019, https://www.verywellfamily.com/how-much-does-ivf-cost-1960212.

14. "New NICE Guidelines for NHS Fertility Treatment," National Health Service, February 20, 2013, https://www.nhs.uk/news/pregnancy-and-child/new-nice-guidelines-for-nhs-fertility-treatment.

15. Myrisha Lewis, "Halted Innovation: The Expansion of Federal Jurisdiction over Medicine and the Human Body," *Utah Law Review* 2018, no. 5 (2018): 1074–1121; Kerry Lynn Macintosh, *Enhanced Beings: Human Germline Modification and the Law* (Cambridge: Cambridge University Press, 2018).

16. Fiona Barry, "US FDA Tweaks Requirements for 12-Year Biologics Exclusivity," BioPharma-Reporter.com, August 5, 2014, https://www.biopharma-reporter.com/Article/2014/08/06/US-FDA-tweaks-requirements-for-12-year-biologics-exclusivity.

17. Michael Mezher, Alexander Gaffney, and Zachary Brennan, "Regulatory Explainer: Everything You Need to Know about FDA's Priority Review Vouchers," Regulatory Affairs Professional Society, December 20, 2019, https://www.raps.org/regulatory-focus/news-articles/2017/12/regulatory-explainer-everything-you-need-to-know-about-fdas-priority-review-vouchers.

18. Organizing Committee for the International Summit on Human Gene Editing, "On Human Gene Editing: International Summit Statement," National Academies of Sciences, Engineering, and Medicine, December 5, 2015, http://www8.nationalacademies.org/onpinews/newsitem.aspx?RecordID=12032015a.

19. National Academies of Sciences, Engineering, and Medicine et al., *Human Genome Editing: Science, Ethics, and Governance*, Washington, DC, 2017.

20. Sheila Jasanoff, J. Benjamin Hurlbut, and Krishanu Saha, "Democratic Governance of Human Germline Genome Editing," *The CRISPR Journal* 2, no. 5 (2019): 266–271, https://doi.org/10.1089/crispr.2019.0047.

21. Eric Lander, Françoise Baylis, Feng Zhang, et al., "Adopt a Moratorium on Heritable Genome Editing," *Nature* 567, no. 7747 (2019): 165–168, https://doi.org/10.1038/d41586-019-00726-5.

22. Victor J. Dzau, Marcia McNutt, and Venki Ramarkrishnan, "Academies Action on Germline Editing," *Nature* 567 (2019): 175. Note an ambiguity in the idea of a consensus—it has largely been discussed as a consensus for going forward with human germline genome editing, but one could as easily put the burden on the other side and demand a broad consensus to stop such activities.

23. Council of Europe, *Convention for the Protection of Human Rights and Dignity of the Human Being with regard to the Application of Biology and Medicine: Convention on Human Rights and Biomedicine* ("Oviedo Convention") (1997).

24. Winston Churchill, Speech in the House of Commons Debate, November 11, 1947, in *Hansard* 444: 206–207. Some have claimed that

Churchill did not originate this statement as he expressly says "it has been said." As far as I can tell, no one has ever found that unnamed earlier speaker though doing so would bring 15 minutes of fame. I strongly suspect Churchill made the speaker up for rhetorical purposes.

Conclusion

1. William Shakespeare, *Macbeth*, act 1, sc. 7.

2. Henry T. Greely, "CRISPR'd Babies: Human Germline Genome Editing in the 'He Jiankui Affair,'" *Journal of Law and the Biosciences* 6, no. 1 (2019): 111–183, 183.

Index

Advisory Committee on
    Developing Global Standards
    for Governance and Oversight
    of Human Genome Editing,
    WHO, 189–190
    registry, 189–190
American Society for Reproductive
    Medicine (ASRM), 266–267
Annas, George, 162
Archaea, 37, 39
Asilomar. *See* Asilomar Conference
Asilomar Conference, 49, 53,
    56–59
    parallels with CRISPR
    discussion, 61–62, 65–66
Asilomar Conference Grounds,
    57–58
Atlantic, The, 110, 157
Autosomal dominant, 226–227.
    See also Autosomal recessive;
    Mendelian genetics
    and gene therapies, 230–231
Autosomal recessive, 226–227,
    239, 273–275, 282. *See*

also Autosomal dominant;
    Mendelian genetics
and gene therapies, 230–231,
    237, 273–275

Bacteria, 9, 33–34, 37–39, 41,
    44–45, 52, 139, 210–211
Bai, Chunli, 113
Baihualin China League, 14
Baltimore, David, 55–57, 61–62,
    65–67
    biography, 56–57
    chair of International Summit
    on Human Gene Editing
    organizing committee,
    66–67, 102, 106, 127,
    181
Barrangou, Rodolphe, 96
Baylis, Françoise, 67
Begley, Sharon, 10–11, 46
Belluck, Pam, 124
Berg, Paul, 50–53, 55, 57, 60–62,
    65–67, 186
    biography, 50–51

ß-globin, 9. *See also* Beta-
thalassemia; Hemoglobin-
beta; Sickle cell disease
Beta-thalassemia, 9, 273,
284. *See also* ß-globin;
Hemoglobin-beta
Biological License Application
(BLA), 81
Blastocysts, 15, 117, 149, 249–
251, 254
Broad Institute, 41, 43–45, 67,
107, 187
Broad societal consensus, 68, 106,
181–183, 187–188, 286–287
Brown, Timothy (also "the Berlin
patient"), 155–156
Budding, 26

Carrier, 274–277
Cas protein, 30, 44, 50, 139. *See
also* CRISPR; Enzyme
Cas1, 39
Cas9, 15, 17, 39–41, 44, 65, 139,
163
Catholic Church, 196, 228, 270
CCR2, 151–152
CCR5. *See also* CD4; HIV; T cell
CCR5Δ32, 152–153, 155–158,
161
and influenza, 106, 152, 222,
238
32-base-pair deletion, 13, 17, 18,
152, 171, 236
and West Nile virus, 152, 238
effects of absence of working
CCR5, 150–159

as example of disease
preventing uses of genome
editing, 236–238
He's edits, x, 12–15, 18, 27–31,
93, 95, 106, 135, 138–139,
144, 161–162, 171
possible brain enhancement
effects of edits, 158–159
Rebrikov's plans, 190–191
CD4, 12, 154–155, 157. *See also*
HIV; T cell
Charo, Alta, 60–62, 66, 68–69,
102–104, 198–199
biography, 69
Charpentier, Emmanuelle, 186
and Jennifer Doudna, 7–8,
40–41, 43, 45–47
Cheng, FeiFei, 12
Chimpanzee, 259–261
Chinese Academy of Sciences, 67,
73, 113, 134
Chinese Clinical Trial Registry
(ChiCTR), 92, 96, 99
Chinese responses to He
experiment, 109–120,
193–200
Church, George, 13, 41, 47,
94–95, 110–112, 120, 126,
160, 196, 213
biography, 111–112
CIOMS guidelines, 76–77
Clendenin, Dan, 295
Cline, Martin, 140, 174
Clinical trial(s), 70–71, 110,
198, 200, 255, 284. *See also*
Preclinical trials

CCR5 and HIV, 14–16, 93, 105,
    107, 138
guidelines and requirements,
    78–82, 149
recommendations for, 122,
    169–170, 181–182, 189, 205,
    248, 262–266, 286
COAs. *See* Crypto-Orwellian
    acronyms
Coercion, 77, 215, 217, 220
implicit, 220
Cohen, Jon, 122
Cold Spring Harbor Laboratory,
    10–11, 107, 128, 150
Collins, Francis, 110–111, 187
Common Rule, 78, 80, 149,
    159–161
Consent, 77–79, 195, 252
  and He experiment, 15, 94, 107,
    136–140, 147, 160–165
Council of Europe 72, 86–87, 191,
    286, 288
Crick, Francis, 49–50, 56
CRISPR
  history of its development,
    34–37, 39–41, 42
  how it works, 33, 37–39, 41–42
  Mitalipov's findings on embryo
    editing, 230–231
  Nobel Prize, 45–47
  nonhuman uses, 62–63, 194
  origin of the term, 34–37
  patents, 43–45
  risks when used to edit embryos,
    149–152
  what it is, 33–34

CRISPR Journal, 96–100, 171
  He Jiankui ethics article, 96–100,
    170–171
Crypto-Orwellian acronyms, 23,
    33
CXCR4, 154, 156–157. *See also*
    CCR5; HIV; X4 HIV-1
Cynomolgus macaque, 259–260

Daley, George, 61–62, 67,
    182–183
Danisco, 39
Davies, Kevin, 96
Davis, Ron, 52, 55
Deem, Michael, 6, 94, 121–122,
    136–143, 164–165, 175
Deutscher Ethikrat, 204, 207
DeWitt, Mark, 121–123, 134, 168
Dickey-Wicker Amendment, 84,
    251
Direct Genomics, 8, 123, 132–133
Disabled, 221–222
Diversity, genomic, 221–222, 253
DNA, 7, 9, 21, 24, 33–40, 45, 56,
    84–85, 97–98, 124, 126, 211,
    240, 264
  and editing, 17–19, 21, 28,
    31–32, 42, 93, 110, 150, 163,
    194, 215, 249–250, 253, 255,
    275–277. *See also* CRISPR;
    Germline genome editing
  and mutation, 13, 18, 226,
    274–277
  and PGD, 226–228, 231–232
  recombinant, 49–55, 57, 59,
    62–63, 186

DNA (cont.)
  and twins (Lulu and Nana),
      17–21, 93, 110, 150, 158,
      161, 203
  Doctors' Trial (also the Nuremberg
      trials), 76, 148
  Dolly (sheep), xi, 294
  Doudna, Jennifer, 6, 8, 10, 11, 40,
      41, 43, 45–47, 60–62, 66–67,
      99, 101–104, 110, 128
  biography, 63–64
Dzau, Victor, 113

Efcavitch, Bill, 132–134, 168
Egg (also oocyte), 25, 26, 28–30,
      32, 49, 85, 133, 192, 203,
      209–210, 226, 229–233, 243,
      251–253, 260, 276. *See also*
      Germ cell(s); Germline
  donation, 251–252
  fertilized, 15, 27, 153, 231
Embryo, 11, 15, 19–20, 28–29,
      79–85, 95, 113, 114, 116,
      119–120, 133, 140, 150–151,
      157–159, 162, 164, 176, 191,
      210–212, 225–233, 235–237,
      248–255, 260, 262, 277–278,
      280, 283
Enhancement, 30, 71, 97, 106,
      158–159, 205, 208, 214, 217,
      223, 225
  problems with, 239–244, 245,
      276
Enzyme, 28, 39, 44, 50, 56
"Essen patient," 156–157
Ethical principles, 19–20, 77, 195

Eugenics, 205, 220, 282
Eukaryote, 41, 45–46
European Convention of Human
      Rights, 86–87
Ex vivo, 24
Ex vivo embryo, 82, 161, 227,
      248–250, 254, 255–256,
      261–262, 288

Facility restrictions, 266–267
Ferrell, Ryan, 16, 93–99, 132,
      134, 142–143, 168. *See also*
      HDMZ
First International Summit on
      Human Genome Editing,
      66–67, 83, 169, 180–181,
      183, 286
Food, Drug, and Cosmetic Act of
      1938 (FDCA), 80
Fourteen-day rule, 254–255
Frameshift, 18. *See also* Mutation
Frankenstein, 294

Gene therapy, 28, 30, 62, 65–66,
      111, 140, 174, 232–233, 235,
      237, 239, 275
Genome Editing: An Ethical Review,
      71, 206
*Genome Editing and Human
      Reproduction*, 71, 206
German Ethics Council. *See*
      Deutscher Ethikrat
Germ cell(s), 24–26, 28–29, 186,
      192, 210. *See also* Egg; Sperm
Germline, 24–27, 29. *See also* Egg;
      Germ cell(s); Sperm

Germline genome editing, 7,
        62–88, 104, 110, 113, 119,
        136, 147, 161, 169, 180–191,
        194, 204–297
    definition, 23–32
God, 217–219
    playing God, 59, 228
Good Clinical Practices, 78
Gordon Conference, 54
Guangdong Province, 5, 7, 18–20,
        115, 120, 130, 141, 165, 167,
        195, 200
Guide RNA, 15, 38, 40, 42. *See
        also* Cas protein; CRISPR

HDMZ, 95
Harmonicare. *See* Shenzhen
        HarMoniCare Women's and
        Children's Hospital
He experiment, 99, 105, 121, 123,
        137, 140, 142, 148–149, 159,
        161, 173, 175, 193, 199, 203
Helicos, 132–133
Helsinki Declaration, 76–77,
        148–149, 161
Hemoglobin-beta, 42. *See also*
        ß-globin
He Jiankui
    and Benjamin Hurlbut, 129–132
    conference participation before
        Nov. 2018, 11–12
    and Craig Mello, 123
    and ethics article, 96–98
    and the ethics of his
        experiment, 147–171
    harm to science, 294

human germline genome
        experiment, ix-x, 15–21
    and Mark DeWitt, 122–123
    and Matt Porteus, 126–128
    and Michael Deem, 6, 94,
        137–141
    nonhuman CRISPR research, 10
    and others who knew his plans,
        132–136
    personal background, 5–8
    preparation for human editing,
        12–15
    revelation of experiment, 91–96,
        98–100
    and Steve Quake, 7, 8, 14–16,
        98–99, 124–126
    after the Summit, 116–120
    at the Summit, 105–108
    before the Summit, in Hong
        Kong, 101–105
    trial, 18–21, 116–119
    and William Hurlbut, 10, 115,
        119–120, 128–129
HIV, 12–20, 93, 95, 100, 103, 107,
        117, 151–159, 162–163, 165–
        167, 170, 171, 191, 195, 236,
        238, 259
    and ART in China, 165–166
Hong Kong Academy of Sciences
        (also Academy of Sciences of
        Hong Kong), x, 27
Hong Kong Summit. *See* Second
        International Summit on
        Human Genome Editing
Human Assisted Reproductive
        Technology Specifications, 88

Human Fertilisation and
Embryology Act of 1990, 85
Human Fertilisation and
Embryology Authority
(HFEA), 73, 85, 170, 207,
219, 267, 279, 283, 285–286
Human Genome Editing
Initiative, 66
Human germline genome
definition, 23–30
nonexistence of, 209–215
Human germline genome editing
health care coverage of, 281–283
possible regulation of, 274–281
safety of, 247–267
Human subjects research, 75–80,
87, 138, 140, 149, 158, 162,
164, 195, 197, 263
Humility, 180–184
Hunan province, 5, 14
Huntington's disease, 213, 227,
279
Hurlbut, Ben, 122, 142, 293
biography, 130–131
Hurlbut, William, 10, 115, 119–
122, 128, 168
biography, 130–131
Hynes, Richard, 68

IACUC. *See* Institutional Animal
Care and Use Committee
ICH. *See* International Council
for Harmonisation of
Technical Requirements for
Pharmaceuticals for Human
Use

ICH-GCP. *See* Good Clinical
Practices
Immune system, 12, 27, 151, 154,
238. *See also* T cell
bacterial, 37–40 (*see also*
CRISPR)
Implantation, 9, 15, 85, 95, 153,
160, 164, 191, 227–228,
254–255
IND. *See* Investigational New
Drug exemption
Induced pluripotent stem cells,
47, 229, 252
Innovative Genomics Institute
(IGI), 61
Institutional Animal Care and
Use Committee, 195–196,
261
Institutional review board, 79–80,
128, 140, 149, 166
International Commission on
the Clinical Use of Human
Germline, 185–186
International consensus, 147,
149, 286
International Council for
Harmonisation of
Technical Requirements for
Pharmaceuticals for Human
Use, 78
International Summit on Human
Genome Editing. *See* First
International Summit on
Human Genome Editing;
Second International Summit
on Human Genome Editing

Investigational New Drug exemption, 80–85
In vitro, 24, 29, 63, 65
In vitro fertilization, 18, 88, 111
In vitro gametogenesis (IVG), 62, 252–253
IRB. *See* Institutional review board

Jansen, Ruud, 34
Jasanoff, Sheila, 131–132
Jasin, Maria, 104–108
Joseph, Andrew, 10–11
Journal of the American Medical Association, 144

Knoepfler, Paul, 231

Lander, Eric, 45–47, 67, 186
Lemley, Mark, 43
Liu, David, 107
Lombardi, Steve, 133–134, 168
"London patient," 155–156
Lovell-Badge, Robin, 67, 99, 100, 102–106, 126, 184
Lulu, 16, 19, 108, 133, 139–140, 144, 153, 155, 157–158, 169. *See also* Nana; Twins

Macbeth, 293
Marchione, Marilynn, 93–95, 135
Mazza, Anne-Marie, 101, 102
McNutt, Marsha, 113
Medical tourism, 133–134, 280
Mello, Craig, 121, 123–124, 133–134, 168

Mendelian genetics, 208, 226, 232, 234–235, 237, 240
Mice, 10, 13, 47, 151, 158–159, 211, 229, 252, 257
Microsoft Word, 41
Mitalipov, Shoukhrat, 230–231
Mitochondria, 30, 84, 231, 274
Mitochondrial transfer, 81, 84, 88, 133, 220, 283
MIT Technology Review, ix, 91, 124
Mojica, Francisco, 34–37, 39, 47
Monitoring, 73, 170, 248, 257, 261, 265, 282
Monkey, 10–11, 24, 101, 151, 258–261
embryos, 13, 101, 138, 143, 151, 159, 194
Moratorium, 53, 84, 185–189, 286–287
Morgan, Julia Hunt, 58
Mosaic, 28–29, 150
Mosaicism, 135–136, 250
Musunuru, Kiran, 11–12, 14, 94–95, 100, 135–136, 157
Mutation, 11, 13, 26, 28, 163, 209–210, 212, 214, 275. *See also* CCR5; Frameshift

Nana, 16, 19, 108, 133, 139–140, 144, 153, 155, 157–158, 169
Nanshan District People's Court of Shenzhen City, 20
Napa meeting, 60–61, 63, 65–67, 110
Naturalness, 205, 208, 217–218, 270–271

Nature (journal), xi, 3, 9, 45, 59,
66, 83, 134, 138–139, 143–
144, 169, 186–192
Chinese ethicists article,
199–200
and the "Essen patient,"
156–157
NDA. *See* New Drug Application
New Drug Application (NDA),
80–81, 284
Niakan, Kathy, 85, 104, 106, 166
Nobel Prize, 45–47, 52–53, 55–56,
61, 64, 112, 123, 125, 186
Nonhuman primate(s), 127, 258–
262. *See also* Chimpanzee;
Cynomolgus macaque;
Rhesus macaque
Non-Mendelian disease, 208,
234–235. *See also* Mendelian
genetics
Non-Mendelian trait, 226. *See also*
Mendelian genetics
Nuffield Council on Bioethics,
71, 198
2016 report, 71
2018 report, 71–73, 75, 104,
169, 170, 181, 204, 206
Nuremberg Code, 76, 147, 149

Off-label use, 284–285
Off-target, 10–11, 114, 150–151,
163, 192
Olson, Steve, 67
Oviedo Convention, 86–88,
207
Oviedo Protocol, 191

Pando, 26
Parthenogenesis, 26–27
Patent, 53, 233, 28, 284
and CRISPR, 43–46, 64
PCSK9, 11–12, 136, 138–139, 143,
159, 236, 238
Peacock Program, 7–8
Pei Duanqing, 134, 168
PGD. *See* Preimplantation genetic
diagnosis
Porteus, Matt, 42, 106–107, 122,
126–130, 134, 142, 168,
178–179
Precautionary principle, 271
Preclinical trials, 248–262
Preimplantation genetic
diagnosis, 163, 208, 215,
223, 227–235, 237, 240, 245,
275–277, 279
Princess Bride, The, 144
*Proceedings of the National Academy
of Sciences*, 40–41, 59
Public health, 238

Qin Jinzhou, 20, 95, 118
Qiu, Jane, 137, 198–200
Quake, Steve, 7–8, 11, 14–16,
98–99, 121, 124–125,
129–130, 132–134, 142,
168
biography, 125–126

Raiders of the Lost Ark, 209
Rebrikov, Denis, 190–193
response of Russian scientists
to, 192

Recombinant Activities
    Committee (RAC), 59
Regalado, Antonio, 3, 88, 91–96,
    100–102, 129, 138, 144,
    158–159, 168
Renzong, Qiu, 199
Reproductive tourism. *See* Medical
    tourism
R5 HIV-1, 154–155, 158
Rhesus macaque, 259–261
Rice University, 5–6, 9, 94, 121,
    136
  investigation into Deem, 137,
    140–143, 175
Rider, 82–84. *See also* Dickey-
    Wicker Amendment
Risk Evaluation and Mitigation
    Strategies (REMS), 279
RNA, 7, 37–41, 55–56, 64, 123,
    249–250, 255. *See also* Guide
    RNA
Rotifer, 26
Ruipeng, Lei, 199

Savulescu, Julian, 125
Science (journal), 9, 14, 40–41,
    45, 54–55, 65, 67, 96, 98,
    102–104, 110–111, 113, 122,
    130, 132–133, 143, 169, 180,
    183, 185–186
Science Translational Medicine, 157
Second International Summit
    on Human Genome Editing
    (also Hong Kong Summit),
    17, 32, 73, 91, 96, 98–108,
    110, 115, 118–119, 126–129,
    134, 136, 142–143, 165, 171,
    181–182, 184
Shenzhen HarMoniCare Women's
    and Children's Hospital, 14,
    94, 166–167
Sherkow, Jacob, 43
Sickle cell disease, 42, 126, 232,
    273, 284
Šikšnys, Virginijus, 40
Silva, Alcino, 159
Snitching, 176–179, 193
Somatic cell(s), 24–27
Somatic gene editing, 28–29, 67,
    70, 81, 186, 189, 203, 205,
    215, 222, 225, 233, 235,
    240–242, 245. *See also* Gene
    therapy; Germline genome
    editing
  cell therapy, 62–63, 66, 232,
    235, 237
Southern University of Science
    and Technology, 8, 14,
    19–20, 92, 103, 115, 130, 166
Species, 242–244
  speciation events, 225, 243–245,
    288, 294
Sperm, 9, 25–30, 32, 49, 85, 133,
    153, 203, 209–210, 226, 229,
    232–233, 243, 251–253, 279.
    See also Egg; Germ cell(s);
    Germline
  sperm washing, 107, 165
Stanford University, 7, 10–11,
    50–53, 129–130, 142, 194
STAT, 92, 102–103, 119, 122–123,
    128–129, 137–140, 143, 198

Sun Yat-sen University, 9
SUSTech. *See* Southern University of Science and Technology

T cell, 12, 18, 151, 154–158. *See also* CCR5; CD4
TALENs. *See* Transcription activator-like effector nucleases
Tam, Patrick, 103
Templeton Foundation, 10
Thousand Talents program, 7–8
Topol, Eric, 94, 110
Transcription activator-like effector nucleases (TALENs), 31, 194
Trial of He Jiankui, 115–118
Twins, 16–18, 93, 95, 98–99, 104, 110, 138, 141, 152, 159, 161, 209, 211, 254. *See also* Lulu; Nana

UNESCO, 72, 76, 209
U.K. Royal Society, x, 66–67, 73, 184–185, 187, 287
U.S. National Academy of Medicine, x, 54, 64, 66–69, 71, 113, 125
U.S. National Academy of Sciences, x, 54–55, 57, 64, 67–69, 71, 73, 113, 125
University of California at Berkeley, 10, 14, 40, 61, 64, 103, 122, 125, 128

Valentine's Day Report, 70–71, 134, 181, 204–205

Wall Street Journal, 16, 59
Watson, James, 49–50, 55
WeChat statement, 114, 199
Wei Zhu, 199
Weismann, August, 24
Wolinetz, Carrie, 187
World Health Organization (WHO), 76–77, 185–186, 188–192, 194

X4 HIV-1, 154–158
Xi Jinping, 198
Xiaomei Zhai, 199
Xinhua (also *XINHUANET*), 3, 18–20, 115–118, 165, 167, 195

Yamanaka, Shinya, 47
Yan Zen, 6
"Yellow Peril," 194
Yong, Ed, 110
YouTube, x, 11, 17, 91, 95, 100, 168, 171

Zayner, Josiah, 119
Zhang Renli, 20, 98, 118
Zhang, Feng, 41, 43, 45–47, 183, 186
Zhang, John, 133–134, 168, 220
Zhitong Lin, 94, 166
Zinc finger nuclease (ZnFs), 31
Zygote, 28–29, 88, 232, 250